Life in Deep Time

Darwin's "Missing" Fossil Record

Life in Deep Time

Darwin's "Missing" Fossil Record

J. William Schopf

CRC Press is an imprint of the
Taylor & Francis Group, an **informa** business

Cover: Precambrian microscopic fossils (*Primaevifilum amoenum*) from the ~3,465-million-year-old Apex chert of northwestern Western Australia.

CRC Press
Taylor & Francis Group
6000 Broken Sound Parkway NW, Suite 300
Boca Raton, FL 33487-2742

Printed on acid-free paper

International Standard Book Number-13: 978-1-138-39037-9 (Hardback)
International Standard Book Number-13: 978-1-138-38549-8 (Paperback)

Visit the Taylor & Francis Web site at
http://www.taylorandfrancis.com

and the CRC Press Web site at
http://www.crcpress.com

To my teachers, from whom I learned,

to my students, who teach me still,

to my wife, who cheers me on,

and to my colleagues in the PPRG

who set this paradigm-changing science on its current course.

Contents

Introduction

INTRO. 1: DARWIN STATES THE PROBLEM

In 1859, Charles Darwin stated the problem: There was no known evidence of life before the oldest animal fossils (dating from roughly 500 million years ago), no evidence of life's earlier evolutionary history. To Darwin, this posed an "inexplicable" gaping hole in his theory of evolution and, for the following century, this "greatest unsolved problem in Natural Science" remained just that – unsolved – the "missing" fossil record of primordial life remaining unknown and assumed unknowable. We now know that the record of life extends to at least 3,500 million years ago, a seven-fold increase since Darwin and an advance spurred by breakthrough discoveries 100 years after Darwin first broached the problem. How and why did this sea-change finally occur? This narrative from the last living participant in the mid-1960s so-called "vanguard" of the handful of workers involved in these paradigm-changing discoveries reveals its inner workings, its whys and wherefores, its ups and downs, the scientists and events that helped and hindered this change in human knowledge, and how the breakthrough finds were accomplished. With this background in hand, the discussion then outlines the present status and future development of this field of knowledge.

INTRO. 2: WHY THIS BOOK?

In his epochal 1859 volume *On the Origin of Species*, Charles Darwin returned repeatedly to what he regarded to be the principal flaw in his theory of evolution – the vexing dilemma posed by the absence of any record of life in rocks deposited before the beginning of the Cambrian Period of Earth history, some 500 million years ago. As Darwin phrased it,

> If the theory [of evolution] be true, it is indisputable that before the lowest Cambrian stratum was deposited long periods elapsed, as long as, or probably far longer than, the whole interval from the Cambrian age to the present day; and that during these vast periods the world swarmed with living creatures. [However], to the question why we do not find rich fossiliferous deposits belonging to these assumed earliest periods prior to the Cambrian system, I can give no satisfactory answer … the difficulty of assigning any good reason for the absence of vast piles of strata rich in fossils beneath the Cambrian system is very great. … The case at present must remain inexplicable; and may be truly urged as a valid argument against the views here entertained.

In Darwin's day, the oldest known fossils were trilobites, advanced lobster-like animals that were far too large, many-celled, and complex to have anything whatever to do with life's earliest stages. Nevertheless, Darwin realized, as did many of his contemporaries, that if his theory were true there simply had to be an enormous amount of evolution – from simple to complex, from single-celled to many-celled – during pre-Cambrian, pre-trilobite "deep time," now known to encompass four billion years, nearly 90% of Earth's existence (Figure I.1).

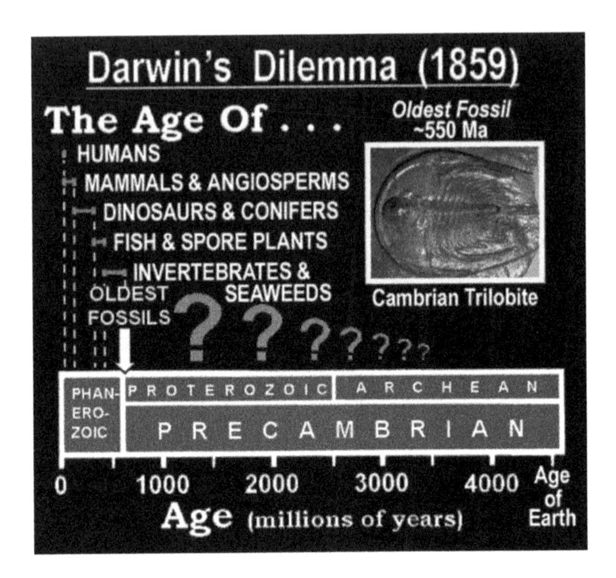

FIGURE I.1 Darwin's dilemma.

To Darwin and the adherents of his theory, evidence of life's early stages was unknown and thought unknowable, an absence that they artfully explained away – perhaps the earlier rocks and the fossil they contained had been wiped from the rock record by erosion, or had been destroyed by geological pressure-cooking, or were still awaiting discovery in some far-off land. Or perhaps the earlier forms of life were too small – or too fragile – or too few to have survived in the rock record. Seemingly plausible and in part correct, such assumptions held sway not only in Darwin's day but also throughout the entire following century – and Darwin's quandary remained unanswered.

Inroads into the solution of this vexing problem were made in the early 1900s – only to be scuttled by influential naysayers, most prominently Sir Albert Charles Seward, the world's leading paleobotanist and vice chancellor (i.e., "CEO") of England's Cambridge University. The problem then lay dormant until the 1950s – but here again, the glimmerings of progress were ignored by the international scientific community, influenced largely by East–West tribalism and the then-raging Communism vs. capitalism Cold War.

Indeed, it was not until 1965 that the two breakthrough publications appeared, both in the "hot-topic" journal *Science*, the first by Elso S. Barghoorn and Stanley A. Tyler describing microscopic fossils discovered in the mid-Precambrian Gunflint chert of Ontario, Canada – a manuscript that, after Tyler's untimely death, I helped Barghoorn prepare, the Gunflint fossils having been the subject of the Honors thesis I had completed two years earlier as an undergraduate at Oberlin College in northeastern Ohio before I journeyed to Harvard to become Barghoorn's student. The second 1965 breakthrough publication – which Barghoorn and I co-authored – reported younger Precambrian microorganisms from the Bitter Springs chert of central Australia, a rich assemblage of more types of appreciably better-preserved early-evolved microbes than those of the Gunflint deposit, a preliminary but convincing

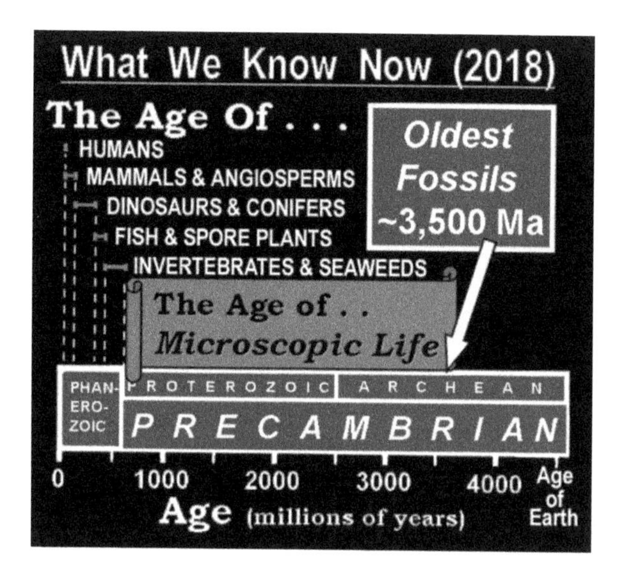

FIGURE I.2 What we know now.

"deal-sealing" report that, together with the previous Gunflint paper and after a century of unrewarded effort, showed once and for all that a solution to Darwin's dilemma actually exists, that the so-called "missing" fossil record of primordial life was in fact not "missing" after all. Since then, thanks to the studies of a great many workers worldwide, the history of life is now known to extend to at least 3,500 million years, a seven-fold increase since Darwin's time (Figure I.2), a change in knowledge over the past 50 years in which I participated, witnessed first-hand, and played what some regard to have been a significant role.

It is undeniable that the 1965 decisive change from the earlier "unknown and unknowable" to firmly established and now universally accepted – proving fruitful only after more than a century of missteps, hopeful but erroneous claims, and valid discoveries that had been cast aside – is an advance in knowledge that University of Chicago philosopher Thomas Kuhn would have pegged a "paradigm shift" in his classic 1962 book on *The Structure of Scientific Revolutions*. And while this particular paradigm shift pales in comparison with Copernicus' realization that the Earth flies around the Sun, not the other way around, it too is fundamental, the root of two new fields of science, Precambrian Paleobiology and Astrobiology, our ongoing search for life on other worlds. In fact, it is the discovery of such ancient Precambrian fossils and the no longer mysterious history of early life on Earth that today provide the basis of NASA's strategy to search for evidence of microbial life on Mars rather than of larger organisms such as snails or worms that would be far more obvious and easier to detect.

INTRO. 3: WHY THIS PERSONAL PRESENTATION?

Though I would prefer not to feel compelled to write this lengthy narrative, I owe it to my field and to the history of science. I was there from the beginnings of the breakthrough advances and know first-hand the whys and wherefores of this change in human understanding. Moreover, I now find myself to be the last living participant in what Preston Cloud, a principal player in these game-changing events, referred to in 1983 as the "vanguard" of the field, the 1960s handful of workers who made the prescient breakthrough discoveries (most prominently, Boris Timofeev in the former Soviet Union; Martin Glaessner in Australia; and Stanley Tyler and Elso Barghoorn in the United States), the founding greats of the modern science. All of them and even a number of my contemporaries and close colleagues (e.g., Hans Hoffman, Canada; and Kase Kline and John Hayes, the United States) are no longer with us. Except for Stanley Tyler, who died during my graduate-school days before I could meet him, I knew them all. They were my teachers and friends, and I feel a personal obligation to them to record the history of this episode in the development of science in which they collectively played such a pivotal role. In short, I have come to realize that unless I record the history of this paradigm change, it will likely be lost, and no one will truly understand how and why it occurred.

I certainly was not the first to highlight the problem of the so-called "missing" record of early life – that was Darwin. And I was not even the one to find the first telling fossil evidence – that prize belongs to Boris Timofeev and Stanley Tyler. But as things have turned out, I have come to realize that because I was evidently the first to see through the problem at a young age, when I was a 2nd-year college student, and because over the nearly 60 years since those days I have persisted in my quest to provide an answer to Darwin's dilemma – as I discovered the first such ancient fossils in several countries and established an international interdisciplinary team of scientists to attack the problem – it is I who has been pegged as leading the charge to forge this field to its present status.

The upshot of this is that a part of this essay is about my experiences and my values – which although no doubt sermonesque in places cannot readily be omitted because I know full well that no one would understand how this game-changer came to pass unless they properly understand me, the "reporting participating witness," and my background, motivation, and perseverance. And though I suppose that each and every such paradigm-shifting advance must have been promoted by some dedicated individual who, for one reason or another, saw through a problem and rallied others in support, it is rare to have a personal account from a motivating linchpin.

At the same time, the obviously necessary personal aspect of this presentation gives me pause because I dislike being out there "front and center" – simply put, I am not a "me-firster." To mitigate this unpleasant prospect I remembered a conversation I had with Tom Kuhn when some years ago he first urged me to write up my account of these game-changing events. To paraphrase his comments, he suggested, "Bill – imagine – we would be far more enlightened about such paradigm shifts had it been possible for a knowledgeable investigative reporter – say, from the *Washington Post* or the *New York Times* – to have quizzed Copernicus or Darwin or a participating witness of the seminal events." Kuhn then went on to list the sorts of questions the reporters might ask:

> How did the game-changers get their ideas, how did they have the courage to carry them through to fruition, and could the imagination and perseverance required to make the breakthroughs have been predicted from their upbringing, their personalities, their

> behavior when they were young? Were their sea-changing concepts original with them or did they evolve from the ideas of others and, if so, from whom, and why did these bits a pieces of the puzzle make a difference to them but not to others of their time? What was their motivation – curiosity, personal values, or a perceived need for fame? How did they manage to weather the storm of doubt, criticism, and contempt that their findings were sure to engender? And how did their breakthrough discoveries affect them personally – were they afraid, or pleased, or did they imagine that their contributions were simply all in a day's work?

Importantly, Kuhn thought, such reporters would seek to understand the underpinnings of the breakthroughs, not just their final results.

In Kuhn's frustration at not having available the data he sought to establish fully the roots of such major changes in knowledge, he concluded his pitch by noting to the effect that

> For Copernicus, the answers to such questions are largely lost to history, whereas Darwin left us his writings, most notably his diaries, that provide appreciably more insight, reinforced by the writings of his colleagues Lyell and Hooker. But the underlying answers to such questions – not just their sea-changing conclusions that the Earth encircles the Sun and that over time life evolved by its adaptation to its changing physical and biological environment – are precisely what are needed by historians and philosophers of science, and all scientists generally to understand the how and why of such paradigm shifts.

As I harked back to this conversation, I came to realize that Tom Kuhn had given me a way out of the conundrum I faced of trying to tell the story that he thought important without presenting myself as some sort of self-serving so-and-so. The story here presented is therefore cast as though it were written by an experienced, knowledgeable investigative reporter who both understands and is privy to my view of the relevant facts (a third-person form of presentation that is maintained until the "Final Comments" of the very last chapter, where I could find no way to properly fully credit others – especially my colleagues in the paradigm-solidifying Precambrian Paleobiology Research Group – without finally switching back to a first-person format).

In short, the answers to Tom Kuhn's queries about the how, why, wherefores, and underpinnings of the solution to the quandary posed by the "inexplicable" absence of a record of the formative stages of life on Earth are here presented.

INTRO. 4: HOW IS THIS STORY STRUCTURED AND WHY?

From the outset of this story, the reader should please understand that my narrative is intended to be a non-technical account, not a formal nitty-gritty, reference-laden scientific presentation. As such, in my effort to make it "accessible" to diverse readers, I am likely to have oversimplified various matters that others would have addressed in more detail. Moreover, parts of the narrative are based solely on my remembrance of events now long passed. Useful as this may be to historians – if my memory has not failed me – it could also introduce unforced errors which, of course, I have tried to guard against. Still, I am not infallible. All I can promise is that I have told the story as honestly as I can and that I have not knowingly misled. Please note also that backing for various assertions in the narrative is provided both by images accompanying several of the chapters and by a fairly extensive "Citation of Selected Relevant Literature" (1796–2018) in the Addendum to the volume.

As for the structure and content of the volume, it is necessarily chronological, first about me – my genealogy, my upbringing, my schooling, my commitment to unearthing the answer to this long-unsolved vexing problem – personal vignettes that will help explain my values, motivation, and role in this science-changing venture that I hope will prove helpful and perhaps even inspiring to students who are seeking their path through life. After all, I began this quest when I was an inexperienced 2nd-year college student and had become curious about what the professor in our class had asserted to be the "Greatest unsolved problem in the Natural Sciences." I was skeptical: I checked – he was correct.

Nevertheless, it made no sense to me that such a fundamental "unknown" had persisted for the entire preceding century. So, as an enthusiastic know-nothing neophyte, I set out to pursue the problem. Repeatedly, well-meaning knowledgeable elders tried to dissuade me, suggesting that given the well-known long history of the problem, my quest seemed certain to end in embarrassingly unbridled failure. But I persisted … and succeeded. And the success came remarkably quickly, far faster than I could have imagined even in my wildest dreams. My message to the students is that there is in fact nothing particularly "special" about me. If I can contribute to such a sea-change – and if the students believe in themselves – they can too.

The structure of my discussion of the development of this science is also chronological – its birth, gestation, and rise to maturity from Darwin and his predecessors in the first half of the nineteenth century, to the paradigm shift in the 1960s, to the present. As the narrative shows, the actual history is appreciably more circuitous than non-scientists might imagine, like all such major advances being a complicated mix of past insights, new discoveries, cultural and political biases, and personality-dependent human interactions, an all-too-typical sinuous path from disbelief to doubt to grudging then widespread acceptance.

A second theme of this story has been my long-term effort to mold this field into an interdisciplinary international endeavor. Today such ventures are fairly commonplace in science, accepted and simply taken for granted. But they were not in the 1970s when I assembled the first such team (which we dubbed the Precambrian Paleobiology Research Group, the PPRG) to work together in my lab at the University of California, Los Angeles (UCLA), for a 14-month period. At the time, I received quite a lot of flak about this notion from skeptical senior scientists who claimed that it simply would not work. As history shows, however, they were mistaken – and the science is now broadly interdisciplinary (appreciably more than virtually any other area of Natural Science) and markedly international. This was not a matter of happenstance. Rather, it was thought through in advance and rather carefully crafted, another lesson about how science actually progresses, in this case illustrated by anecdotes from my various research visits to Australia, China, India, and the former Soviet Union, where I discovered the first three-dimensionally well-preserved Precambrian fossils there known.

You may wonder, quite reasonably, "Why my penchant for interdisciplinary internationalism?" The interdisciplinary emphasis comes from my education – in

Geology as an undergraduate, and in Biology as a graduate student – schooling that led me to understand that the questions I wanted to answer would require a mix of these broad fields and of Microbiology, Organic Chemistry and Biochemistry, Organic and Isotopic Geochemistry, Paleontology, Sedimentology and Comparative Planetology, for me backed by logic-based schooling in Philosophy and linked together by Darwinian evolution. I imagined the area of science I wanted to build to be like the Dixieland bands I had headed in high school and college – one member "singing" on a trumpet, another on a trombone, a saxophonist blaring forth, a pianist plunking the keys, yet another strumming the base, all held together by a drummer "beating the skins." Each excellent, each a master of a specialized craft, and all bonding together to make music that none could accomplish alone.

My commitment to internationalism stems from my upbringing, the values embedded in me by my parents during my formative years – but as I learned years later, not just from them but from my ancestors' value-laden Mennonite tradition, reinforced in me by my experiences and adventures in many countries. To wit, I then, now, and will always believe that nationalistic "tribalism" is detrimental to human interactions and the improvement of society, not only in the United States but for all human societies, with such tribalism being particularly detrimental to the progress of science. Indeed, it seems to me that national origin, religion, gender, politics, and the like should play no role at all in science. There is only one reality. Science's task is to unlock that reality and describe it and understand it as best it can. And no particular group – not one of the numerous and disparate human "tribes" – holds all the keys.

INTRO. 5: TAKE-HOME LESSONS

As you read this work, you will learn about how this remarkable paradigm shift arose and the fitful starts, lags, and roadblocks during its century-long gestation; about how the seminal discoveries of the 1960s then spurred the science and how, with the impetus provided by the PPRG, it has since matured with new findings and new techniques; and about where this science is now and the major questions it yet needs to address. And not only will you learn about how this science works and the reasons for its remarkable advance but also about its occasional undermining by cultural biases, "me-first" irrationality, and their all-too-common hurtful by-product, "professional jealousy."

A take-home lesson I hope to convey is that science, like whichever society the science is embedded in, moves in fits and starts, three steps forward then two steps back – the gradually accumulating forward steps ultimately adding up – and that scientists are human, subject to the same strengths and foibles as all others.

Such traits are particularly well illustrated in the section of Chapter 8 that outlines the saga of a decade-long disagreement over the Earth's oldest records of life. As that story not surprisingly demonstrates, the humanness of scientists comes to the fore in such internecine battles, when it can sometimes spill over into *ad hominem* attacks and confusion of an unwelcome message and the bearer of that message. Fortunately, however – as that episode also illustrates – one of the great glories of science is that it is self-correcting, embodying within it the wherewithal to cast aside irrational or unfounded notions. In short, what matters in science is whether the facts hold up and whether their suggested interpretation fits the evidence at hand and makes good sense. "At the end of the day" (a Britishism to which some object), *facts will always win*.

A broader take-home lesson relates to how paradigm shifts that correct mistaken yet long-accepted dogma actually occur. Though such monumental sea-changes are frequently attributed to the "Eureka Moment" of a single individual – as is often assumed for the *Copernican Revolution* and *Darwin's Evolution* – such breakthroughs are far more commonly rooted in the insights of others who came before. For example, the paradigm-shattering idea of the Renaissance-era mathematician and astronomer Nicolaus Copernicus that Earth encircles the Sun – rather than the Earth being the immobile center of the Universe – actually dates to the Greek astronomer and mathematician Aristarchus (310–230 BC), as Copernicus clearly understood. And even Darwin's ideas were in part derived – from his physician grandfather, Erasmus Darwin; from the French naturalists Jean-Baptiste Lamark and Georges Cuvier; and most prominently from the writings of Thomas Robert Malthus – a history summarized in Chapter 3.

Yet it was Copernicus and Darwin who garnered the credit – which they rightly deserve – because it was their concepts, not those of their predecessors, that ultimately "won the day." And why did they win? It is simply because Copernicus and Darwin had each ferreted out the telling facts to support their dogma-challenging notions. In other words, their concepts were not simply a set of speculative, if assuredly "interesting" ideas. In this sense, at least, it is not Copernicus and Darwin who "won" but, rather, it is the facts they presented. And this, in turn, is also not at all surprising. After all, there is only one reality, and by their fact-supported insights, Copernicus and Darwin each helped us all to better understand the world about us.

The narrative in this book is an attempt to address the long-term need for an explanation as to how Darwin's quandary about what he termed the "inexplicable" absence of a pre-Cambrian pre-trilobite deep-time record of life was finally unearthed, a story that fits well with the historical antecedents of both the Copernican and Darwinian revolutions. Not one of us to whom aspects of this latest paradigm shift may currently be attributed stands alone. In fact, to me the great "Hero" of the sea-change in knowledge recounted here was the early-1900s American paleontologist Charles Doolittle Walcott. Yes, as you will see in Chapter 4, Walcott made errors, and his finds of pre-Cambrian, pre-trilobite "algal" fossils and the mound-shaped structures they constructed ("stromatolites") were in his time not accepted. Still, given the then-prevailing dogma, he was remarkably astute. He was asking the right questions. He was on the right track. And though some of his interpretations missed the mark, many of his discoveries and then-novel ideas have since proven to be absolutely correct. From my view of how this science has advanced, it seems to me that all of us who have contributed to this field and its paradigm-shifting prominence stand on the broad shoulders of C.D. Walcott.

INTRO. 6: CLOSING COMMENTS

In 1854, Louis Pasteur suggested that "*chance favors only the prepared mind.*" This aphorism has applied to my life in science, but only in limited measure.

Indeed, I have had a huge number of "breaks," most of which were well beyond my control – wonderfully nice things for which I was the thankful, if not particularly deserving, beneficiary – and not only in science and academia but in my personal life as well. Simply put, I have been terrifically lucky. I am grateful.

From my perspective, as the now last survivor of what Preston Cloud in 1983 referred to as the paradigm-shifting vanguard – kind of like being "The Last of the Mohicans," the last members of a dying Native American tribe depicted in James Fenimore Cooper's classic 1826 novel – I have no choice except to have written this extended essay. As you will discover, though I am a student of the past, an erstwhile paleobiological historian of deep time, I also care about the future. Now aged 76, in my 57th year in this ongoing quest to uncover life's ancient history and my 50th as an active UCLA professor, "I've been there, done that." It is for those reasons that this work is written for students, paleontologists, microbiologists, evolutionary biologists, historians and philosophers of science, interested laypersons, and anyone else (even our elected officials) who might wonder how such sea-changes actually occur and how science really works. In my effort here to satisfy Tom Kuhn's heartfelt urging of years ago, this book will tell you the how, when, and why we now know that Darwin's quandary about the so-called "missing" record of life's early history has been laid to rest, that life on our planet dates from at least three-and-a-half billion years ago.

INTRO. 7: ACKNOWLEDGMENTS

I thank plant biologist Jane Shen-Miller Schopf, my wife, whose helpful comments on a draft of this narrative have markedly improved its presentation. I also thank my many friends and colleagues, most particularly those of the PPRG, who have advanced this science by "articulating" (in Kuhnian phraseology) this paradigm shift over the past many decades and brought the field to its current status. And on behalf of us all, I thank those who came before – most particularly Charles Doolittle Walcott, who had the prescience to see the problem clearly and the courage to show the science the path to success.

J. William Schopf
January 2018

1 The Schopf Lineage

1.1 OVERVIEW

The story presented in this book is centered on the discovery of the long-sought but long-assumed unknowable "missing" fossil record of the earliest 85% of the history of life, presented from the viewpoint of J. William (Bill) Schopf, a UCLA professor who, for the past nearly 60 years, has been a prime participant in unearthing evidence of the antiquity and development of Earth's earliest biosphere.

1.2 CHRISTIAN SCHOPF

The first Schopf to arrive in this country, in 1739 on the sailing ship *Samuel*, was Christian Schopf from the Rhineland, a loosely defined region embracing the land on either bank of the River Rhine in central Europe. He was a Protestant Calvinist Huguenot escaping religious persecution. At that time, the Huguenots, followers of the Swiss intellectual and political leader Besançon Hughes, were being driven out of Europe by the all-powerful Roman Catholic church which was at odds with their Anabaptist view that baptism and church membership should be reserved only for adults, rather than bestowed immediately after birth, and their commitment to pacifism and the separation of church and state – all of which were regarded as threats to aspects of Catholicism.

Soon after his arrival, Christian Schopf settled in Lancaster, Pennsylvania, where he became a member of the Mennonite Church. This Dutch-derived pacifist Anabaptist sect was one of several such Protestant groups attracted to the New World through the efforts of William Penn, the pacifist Quaker founder of the English Province of Pennsylvania who had toured parts of Europe to entice religious outcasts in an effort to expand the population of Pennsylvania so that it would not be subsumed into the adjacent New York Province. Via various subsequent Schopf families, the clan then moved from Pennsylvania to Ohio, then Indiana, then Missouri, and, ultimately, to the small town of Pine Bluffs in southeastern Wyoming.

1.3 BILL'S FATHER'S FAMILY

Bill Schopf's father's father, Bill's paternal grandfather, was Ira Schopf, and his wife was Nellie Bufton Schopf. Ira was a successful businessman in the small Wyoming town of Pine Bluffs – he had a ranch where Bill's dad rode his favorite pony, and Grandpa Ira owned the one grocery store in town (with an imposing "Schopfs" sign on the side of the building). For many years, Ira was also the mayor of Pine Bluffs, where he lived in the largest, most handsome house in town (which is still extant, as is Bill's dad's one-room school house ... both of which, along with the grave sites of Bill's grandparents, he and his wife Jane visited some years ago during a weekend stop at Pine Bluffs between lectures Bill gave at the Universities of Colorado and Wyoming). By all accounts, Grandpa Ira was an upright citizen and a generous man. During the Great Depression, for example, he is said to have allowed many of the local townspeople to obtain groceries at no cost, knowing (from Ira's entrenched trusting Mennonite values) that they would pay him back when things got better.

After skipping several grades in Pine Bluffs' one-room schoolhouse, Bill's father, James Morton Schopf, graduated from high school at the age of about 13. Because he was too young to enter college, and because his parents were relatively wealthy, he was sent after high school for a two-year stint at Kemper Military Academy in Boonville, Missouri (the town where his father, Ira, had been brought up before moving to Wyoming). After Bill's father died, he read his dad's diary. His dad hated the highly regimented military academy (he and his pals from time to time going "over the wall," AWOL [absent without leave], only occasionally getting caught) and he longed for his Wyoming pony.

Bill's father had two siblings – a brother, Bill (the source of J. William's middle name) who was a year or so younger than he; and a sister, June, two or three years younger still. Bill's uncle died of a stroke, evidently when he was in his 20s, and Bill's aunt June married John Stricklin, a geologic mapper for the US Geological Survey. June raised sheep on a ranch at Roswell, New Mexico. She had a flock of around 40 when Bill visited at age 13, each of which she had named after folks known to her – including Jim (Bill's dad, her older brother), sheep Billy and Tommy (for young Bill and his brother Tom), and a ewe named Maybell (who was about to lamb when Bill visited). Though Aunt June was plenty smart, she had an odd streak – she "read palms," was an adherent of phrenology, and readily accepted the "truth" of the supposed 1950s visit to Roswell by extraterrestrials. Throughout her life, she was annoyed at her older brother, Bill's father, for having convinced their parents that in high school she should learn Latin rather than Spanish, knowledge of which would have much better have served her after she moved to Roswell.

Years earlier, when little Bill was perhaps age five, his dad, accompanied by Bill's older brother, Tom, attended grandfather Ira's funeral (to Bill's knowledge, the only time that his father had ever returned to his Wyoming roots). On his way back to Pittsburgh (to which the family had moved in 1942), he had an automobile accident; the resulting fire incinerated all of the family records that he had acquired.

Bill knows only two other things about his paternal grandparents: (1) Ira and Nellie "lived in sin" for eight to ten years before they married, because Ira had promised his parents that he would marry only a Mennonite, which she was not. When Grandfather Ira's parents died, he finally allowed himself to marry Nellie. (2) As described to Bill by his mother, Grandmother Nellie was a "holy terror." When Bill's mother met her for the first (and only) time, his mother said, "I am pleased to meet you, Mrs. Schaaf" – an incorrect pronunciation of her married name that Bill's dad had permitted for the previous four years – whereupon grandmother Nellie is said to have vehemently replied: "The name is Schopf, it rhymes with 'up' and the 'f' is silent" (because she objected to the spittle that emerged from the final "f"). As one might imagine, Bill's mother was humiliated. Grandmother Nellie is said to have been a proud member of the staunchly conservative "DAR," the Daughters of the American Revolution (for which she presumably qualified by her marriage to Ira, whose roots traced back to Christian Schopf's arrival in 1739).

1.4 BILL'S MOTHER'S FAMILY

On the maternal side of Bill's family, his grandfather was Jorgen Nissen of Cedar Falls, Iowa. Both he and his wife were immigrants from Denmark and, given their Danish heritage, were Protestant Lutheran anti-Catholics – a still extant tribalistic North–South Protestant–Catholic European division dating from the time of Martin Luther in the 1500s. Grandpa Jorgen had a stroke in his later years, and the only time Bill met him he was bed-ridden as he had been for the previous many years, apparently at great expense. Bill met his Grandma Nissen only once and doesn't recall her first name … he knew her only as "Grandma" (but she baked the tastiest muffins Bill had ever eaten!). When their older daughter, Clara, became married, the Nissens gave her half of their extensive and highly productive farmland. They used the other half to fund Bill's mother's college education at Iowa State Teacher's College (now Iowa Northern University).

Bill's mom's sister, Clara (to Bill, "Aunt Sis") was a truly intelligent, decent person. For many years, she hosted a talk show on the local Cedar Falls radio station and was an accomplished pianist. When Bill was 12, he took a Greyhound bus (alone) from the family home in Columbus, Ohio, to Cedar Falls and spent six weeks at their farm, where he worked in the corn fields, shoveled corn and manure, and learned how to milk a cow (the key is to remain quiet and be soft and gentle to the teats); and, since Bill had brought along his trumpet, his Aunt Sis and he played duets.

Aunt Sis' husband (Alfred Hesse, to Bill, "Uncle Alfred") was a braggadocio show-off. (Bill's mother later told him that he'd been that way since grade-school and that Aunt Sis had married him in the hope that he would prove to be a "late bloomer." It never happened!) But Uncle Alfred was popular with his pals in Cedar Falls. When he took Bill in for a haircut, he got the barber to pretend to shave off Bill's "peach fuzz" (using the reverse side of a straight razor). Uncle Alfred thought that was hilarious.

Unbeknownst to Bill, Uncle Alfred was resentful of Bill's parents – simply because they had received an advanced education and he had not. Indeed, Uncle Alfred proclaimed that his "sixth-grade education is all that anyone ever needs!" … a point of view that may have been common for Mid-Westerners of his generation at a time when families were large and the youngsters were needed to work the farm fields. And he had serious misgivings about Bill – not only because, given Bill's parents, it was predictable that this youngster would become educated, but also because young Billy was an inexperienced mere "*city boy.*"

1.5 CITY BOY CHICKEN CHASE

One afternoon, after Bill had finished working in the fields, Uncle Alfred told him "Billy, go down to the chicken coop and fetch us a chicken for dinner." Bill did as he was told. Though he had never been inside the chicken coop, he knew where it was.

Bill went over, opened the door, peered in, and ventured inside. There must have been at least 80 birds perched quietly on their tiered chicken roosts. Bill located a plump one, quietly ambled over, and got a good hold of the chicken's leg. Turmoil ensued. The chicken started furiously flapping; the other birds went berserk; and the coop immediately filled with a huge all-enveloping cloud of sawdust, feed, and bird droppings. Bill let go of the leg, by which time all the other birds had frantically fled the coop and were scurrying around in the chicken yard.

Well, that didn't work! What to do? Bill exited the door, walked around the coop, and went into the yard. Next notion: "Run'em down!" A not unreasonable idea – in football he was a halfback – he was quick, fast, and could cut on a dime. (During the football season later that year, he scored 22 touchdowns in his local youth league.) "Heck" – he thought – "no problem, I can do this." *Not so!* Have you ever tried to chase down a chicken? Gee Whiz are they quick! (Think about it … how else can they elude the chicken hawks? Required for their survival, their ability for lightning-quick zigzag cut-backs is embedded in their genes.) Bill tried repeatedly, five or six times – thought he had 'em only to miss time after time. He glanced back at the farmhouse where his uncle was intently staring out a window, pounding his belly and laughing uproariously.

OK … that also did not work. Twelve-year-old "city boy" was failing. Ahh … but even then, that was not Bill's style. In hopes of figuring out a solution to this dilemma, Bill surveyed the yard and spotted a large roll of baling wire at one corner. Slowly, carefully, with arms outstretched, he walked toward it, with three chickens scurrying in front. Two peeled off, but the third found refuge within the coil. "Ha!" … Bill thought … "I got it!" He reached under the coiled wire, this time managing to capture both legs, and withdrew the bird. The other chickens, now at the far end of the yard, were flapping and squawking furiously – but Bill paid no heed. He'd done his job. Bill delivered his prize back to the farmhouse and gave it to Aunt Sis. Uncle Alfred said not a word.

1.6 BILL'S MOTHER

After graduating from Iowa State Teacher's College, Bill's mom, née Esther Julie Nissen, hoped to do graduate work in mathematics – but when she could not land a scholarship to continue those studies, she switched to botany and earned a Master's degree from the University of Illinois (where she met Bill's father). She then transferred to the University of Michigan, where she completed the requirements to be advanced to candidacy for the Ph.D. (which she never completed), thus earning what is now known as a Cand. Phil. Degree; got married to Bill's father; and returned to Urbana, Illinois, where he had a research position at the Illinois Geological Survey. During Bill's mother's sojourn in Michigan, his father wooed her by "riding the rails" between Urbana and Ann Arbor (like a hobo, hitching a ride by tumbling onto wooden platforms suspended beneath the passing freight train cars along with dozens of others during this time of the still lingering Great Depression).

In 1939, Bill's brother Tom was born, and then Bill himself in 1941. In 1942, their family moved from Urbana to Pittsburgh, Pennsylvania, where Bill's father had a position in the Bureau of Mines; and later, in 1949, to Columbus, Ohio, where he set up a coal geology laboratory for the US Geological Survey and was professor of Geology and Paleobotany at Ohio State University.

In their family, Bill's mom was his "protector." Bill's older brother, Tom – whom Bill revered – pummeled Bill at every opportunity (an unfortunate but understandable product of sibling rivalry), and Bill's father, consistently enamored with Tom, his first-born son, regarded Bill as a rather bothersome afterthought. Time and again, Bill's mom came to his rescue: "Billy is not all that bad." She was a huge help. She showed Bill how to set the table for dinner and introduced him to cooking – for Bill, in later life, it would have been fun to have been a short-order cook – and every night while Bill (of course, not Tom) dried the dishes, she quizzed him on the multiplication tables (multiplication by nine was his toughest to master).

She stuck it out as a "stay-at-home mom" until Bill finished grade-school. She then enrolled at Ohio State (as a so-called "older student"), retook three or four advanced math classes, and landed a job at the Battelle Memorial Institute, a research and development organization. Later, as an Information Specialist, she took courses in Chinese and Russian (in order to translate terms in the one dictionary then available to decipher Russian engineering jargon and rocket codes) and was one of a select few at Battelle invited to view the first Moon landing, to which her work, and that of the engineers to whom she provided Russian and Chinese data, had contributed. She had top-secret clearance with the Federal Bureau of Investigation (FBI) and much of her work was classified. So, much to the annoyance of Bill's father, she would never reveal to the family what project she was working on (except when they had a guest at dinner, when she would divulge a bare minimum "to be polite"). When Bill was in high school, his father once remarked to him that "your mom is smarter than I am" – though, honestly, Bill didn't think he ever really believed that.

After Bill's father died from a rare form of leukemia at the age of 67, and then his brother (from heart failure) only two or three years later at the age of 42, while he was leading a class field trip from the University of Chicago where he was a professor of Geology and Paleontology, Bill's mom became enamored with an old family friend from their Pittsburg days, who was an executive of the U.S. Steel Corporation. His wife had died and he was looking for a mate. According to Bill's mom, his primary goal was to leave each of his offspring an inheritance of one million dollars. He liked to play the card game bridge, so Bill bought his mom a computer and introduced her to a program that would teach her the game. The two had dates, some lasting three or four days, but she lost out. From that time on, she gave up on him and on life generally.

1.7 BILL'S MOM MEETS TV STAR PETER FALK

Earlier, when Bill was singled out to be that year's UCLA All-Campus Faculty Research Lecturer, he had brought his mother out and introduced her to the crowd as "my Mom, all the way from Columbus, Ohio!" Cacophonous cheers resounded through the hall! Then, a few years later, when Bill established CSEOL (the Center for the Study of Evolution and the Origin of Life), his mom came out again. After the center's official "Grand Opening," Bill and his wife hosted his mom and their long-time friends, writer Irving Stone and his editor-wife Jean, to a fine dinner at a newly established Wolfgang Puck restaurant. Peter Falk was at the adjacent table, and Bill's mom, knowing Falk from the TV program *Columbo*, wanted to meet him. But Bill knew the rules – in the Mid-West, at least, the man has to come to be introduced to the woman, not the other way around. What could he do? He wanted to please his mother, but he was hesitant to bother Mr. Falk. Ultimately, Bill acquiesced. He ventured over to the Falk table and introduced himself. Falk quizzed him. "What do you do at UCLA?" Bill told him. Falk responded: "I just read an article about that on the front page of the *Los Angeles Times*." Bill said: "That's my work." Falk then said "Well, in that case, I'll have to meet your mom!" *Bill lucked out!*

As the two walked from Falk's table to Bill's, it occurred to Bill that to avoid embarrassment, he had better tell Peter Falk that the Stones were also at the table. Falk's response: "Great! I've always wanted to meet Irving Stone!" Introductions were made, and Mr. "Columbo" Falk was wonderfully kind, moving about the table and shaking hands with all present. After he had departed, Irving privately asked Bill: "Who was that guy? He called me Irving!" Bill explained. (Irving, though a long-time member of the Academy of Motion Pictures Arts and Sciences, had never heard of Peter Falk.)

1.8 BILL'S FATHER

As things were handled in Bill's family, his dad was the leader, the "boss," in essence the family's "lord and master." Evidently, sometime during his youth, Bill's dad had become estranged from his parents – or so Bill surmised. And Bill imagines that the root of this estrangement was that his dad's father – Bill's grandfather, Ira – was a devout Mennonite, whereas Bill's father was a confirmed atheist, a view of which Ira would have deeply disapproved.

In any case, Bill does not recall his father ever speaking about his parents or his views of religion. At the same time, Bill is pretty sure that this was not the case for the family lore handed down to Bill's brother Tom (two years older than he), because when Tom was in junior high and high school, virtually every evening after the family dinner, he and their dad would retire to their father's study for extended one-on-one talks. This went on for five or six years. Being younger, Bill was always excluded because he "would not be interested in the grown-up things" they discussed. And Tom never, ever, divulged to Bill what they talked about, claiming it to be "privileged information."

Although Bill did not understand it at the time, it now seems obvious to him that these one-on-one sessions with Tom were intended by their father to instill in his first-born son a set of values that he had found personally important and to ensure that Tom would benefit from his dad's life-experiences.

So, for example, because their dad had skipped four years during his schooling (in the one-room school house at Pine Bluffs) – and because their mom had skipped three (a fact that confirmed to them both the usefulness of skipping grades) – Tom graduated from high school a year early and headed off to Oberlin College in northeastern Ohio. Unfortunately, however, because of Tom's immaturity, this very nearly derailed his future prospects. Though Bill had assumed that Tom was doing excellently in college – as he had in virtually everything he had ever tried – Tom discovered early on that it was difficult to cope, that he was in way over his head relative to the other students. In retrospect, this is not all that surprising. At that time, Oberlin was ranked by the *Chicago Tribune*, the ruling gold standard about such matters, as the top undergraduate liberal arts small college in the United States and arguably the world, its student body composed of well more than 50% high-school valedictorians.

The influence on Tom of their demanding, domineering father was pervasive. Because their dad had been so much younger than the others in his classes at Pine Bluffs and had therefore never been successful in sports, Tom was urged to be an athlete (at which he excelled). Their dad was a poor public speaker, so Tom was encouraged to join the high-school debate team (where he again did quite well). Because their dad had been a high-school musician, Tom was forced to learn the accordion (which he hated and wasn't much good at). And because their dad was not at all skilled in interacting with others, particularly people in positions of authority, Tom was coerced into joining DeMolay, the junior branch of the Masons (a chore that Tom much disliked but managed to tolerate for some two years).

Dating from his early childhood, Bill, as the second-born, was pretty much left on his own – a situation that at the time he resented but might better have embraced since in retrospect, he now knows that the self-reliance it fostered in him has turned out to be a benefit. Nevertheless, throughout Bill's youth, Tom was his hero, and Bill did all in his power to help him. He even physically fought to protect him – in the 2nd grade he picked up a slab of cement and tried to "bean" Johnny Kramer, the youngster his brother was fighting (though the combatants then rolled over and the slab landed on Tom's head instead, which ended the fight).

Later that year – after Tom's 4th-grade teacher had prohibited him from playing with his classmates because of his disruptive belligerence – Tom found himself isolated, alone. He asked Bill if he could instead play with Bill's friends. Of course! The two brothers and Bill's pals played together – until Tom plunged a sharpened stick into Bill's upper lip from which blood poured out (its scar is still evident to this day). The brothers went home. As their alarmed mother worked to staunch the bleeding, Bill lied for Tom to protect him from the wrath of their dad (their mom, concerned only about the spurting blood, didn't care at all about Bill's implausible explanation).

Pretty clearly, Bill's older brother Tom was under enormous pressure from their father, beginning even when the two of them were grade-school youngsters.

As noted above, Bill was never privy to the details of his father's upbringing. A non-assertive confirmed atheist, he never spoke about religion – regarded by most cohorts of his generation to be the cornerstone of their value system – noting once to Bill only that when he had reached the proper age, he had been baptized and that, in his words: "As best I can tell, it never really hurt me." To Bill's knowledge, his father attended church only twice – when Tom got married and then at Bill's wedding – and he never came to hear Bill sing solos at Sunday church services, not even on Easter Sunday. But this was not atypical – unlike his attendance at Tom's "performances," he never took the time to see Bill play football or basketball either, came only once to one of Bill's baseball games (a City Championship game, when he showed up in time to see only the last inning), and was never on hand even to see Bill perform numerous times as the "featured" high-school trumpet soloist. For "old Jim" (as his colleagues at Ohio State referred to him), that was simply the way he was.

Over the years, due to their dad's influence, Bill's brother Tom (an outstanding professor at the University of Chicago and founder of the journal *Paleobiology*) consistently "shot himself in the foot" by taking their dad's values to an extreme, unwaveringly telling his colleagues both publicly and privately his unvarnished opinions about their foibles. Though Bill cannot recall Tom having ever been markedly mistaken – and Bill knows that Tom never meant ill will – because of Tom's outspoken honesty, he was far too often perceived as being unnecessarily unkind to others. For a period of a dozen or so years after Bill had joined Tom on the national paleontological scene, colleagues in their shared community privately referred to Tom as the "bad brother" and to Bill as the "good brother."

Beginning from even an early age, Tom was under enormous pressure to please their father. He weathered that ordeal, much to his credit – but it was a pressure that Bill, because he was not particularly important to his father, largely escaped. *Bill was the lucky one!*

1.9 FAMILY VALUES

From time to time over many years, Bill had wondered about the roots of the values that guided his father's life and, even more so, those of his older brother … values that, of course, rubbed off on him, too. Ultimately, Bill discovered the answer.

Bill had lectured at Virginia Tech. He and his wife Jane had three days before they were to be in Philadelphia for the spring meeting of the American Philosophical Society. They drove up the Blue Ridge Parkway through the Appalachians and across western Pennsylvania, where they stopped at Lancaster and visited the Mennonite archives. Astoundingly to Bill, the entire pre-Wyoming history of the Schopf clan was there recorded. Most importantly, Bill became educated about Mennonites and, for the first time, realized that (except for its theism and total pacifism) the life-guiding values of his father, his brother – and of course him, too – were rooted deeply in the Mennonite tradition.

1.10 TAKE-HOME LESSONS FROM THESE EARLY YEARS

The "City Boy Chicken Chase" taught Bill four lessons about problem-solving: (1) Try the obvious solution first. (2) Failing that, try a second notion. (3) Failing that, try a third. And (4) if you have repeatedly failed, it is either because you do not understand the problem (which in this case Bill initially certainly had not), or you understand the problem but you are asking the wrong questions (here Bill repeatedly failed but ultimately figured out the right question).

In science, the former mistake, to not understand a problem, can be remedied – the relevant data are likely to be found in the literature. But the latter error, asking the wrong question to address a problem, is your own darn fault. Ever wonder why it is that when you misplace your eyeglasses you always find them in the last place you look? Ha! You'd been asking the wrong questions! Two other little lessons are illustrated by the Chicken Chase, the first being that you learn more from your errors than from your successes – the mistakes providing a basis for improvement. And second, try mightily to never give up – if the task is worth doing, if the problem is worth solving, keep at it and do it as best you can!

Why does this narrative tell you about Bill's two aunts, the Cedar Falls farm-wife radio talk-show pianist and the Roswell sheep-ranch palm-reader myth-believer? The answer is simple – to Bill, his knowledge and appreciation of them has been of enormous value to him over his life. The fact is that Bill has lived his life in a cloistered environment – when he was growing up, in college, in graduate school, then over his professional career, virtually everyone he has known has been serious and highly educated, a member of his own peculiar "tribe." However – and of importance to him – thanks to his aunts, he has learned to understand that privileged academic "elites" such as he are not mainstream members of American society. Unlike those of his academic ivory tower crowd, Bill's two aunts represent standard, admirable "Middle America." Time and again, he has asked himself "What would Aunts Sis and June think?" And time and again, this has helped him to better understand the aspirations of his fellow citizens and to realize his true place in American society as a member of a scholarly, well-meaning but numerically minor outlier population, not infrequently accused of harboring "hidden knowledge."

Finally, Bill's brief visit to Pennsylvania Dutch (Amish, Mennonite) country provided him with four personally important new insights:

1. His father had gained independence from his family by casting aside what he regarded to be the mystical "God part" of the Mennonite tradition – a decision, Bill imagines, that resulted in him being a family outcast – but his early-learned values were exceedingly deeply entrenched.
2. Without acknowledging the source of his value system – even to himself, Bill imagines – his father handed down these values to his two sons, with Bill's brother Tom adopting them in an extreme fashion.
3. Thanks to this half-day excursion, Bill now understands why it is that, throughout his life, he has aspired to be a "group person" rather than a "me-firster," Mennonites being active worldwide in the remediation of poverty, disaster relief, the economic development of those in need, and the promotion of peace.
4. And at long last, Bill began to understand why it is that even as a grade-school youngster, he did "not compute," his basic values being derived from those of a small religious minority rather than a dominant more prominent sect of the White-Anglo-Saxon-Protestant ("WASP") tradition. It is presumably at least in part for this reason that Bill has had the courage to be different from others, to branch out on his own, to try things that others shy away from, and to tackle long-known scientific problems for which the odds of success were universally regarded as vanishingly low.

A concluding note: A few years ago, one of Bill's department colleagues, evidently frustrated by Bill's day-in day-out up-beat non-combative demeanor, angrily asked him: "Schopf, why do you always take the high road?" Keeping his "cool," Bill did not reply. Heavens, Bill, given his values, was simply trying to do what seemed to him appropriate. Perhaps that colleague will now understand.

2 Bill Schopf's Schooling and First Year of College

2.1 GRADE-SCHOOL

Soon after Bill Schopf was born in 1941, in Urbana, Illinois, where his father was finishing his Ph.D., the family moved to Pittsburg, where he attended Sterrett Elementary School for the first two years. Good stuff! (Billy met his first girlfriend – Joan Store – and invited her for an ice cream cone. Billy didn't have any money – neither Tom nor he ever got an "allowance." So, on this "first date." the youngsters' mothers joined them at the local ice cream parlor and Billy's mom paid. Free ice cream! Excellent! Good stuff!)

In 1949, their family moved to Columbus, Ohio, where Billy finished grade-school at Clinton Elementary School. Billy didn't get into much trouble; had lots of friends; got good grades; began his trumpet playing; sang in the choir of the local Methodist church (including solos on Easter Sunday); carried newspapers, first in the afternoon, then later in the afternoon and morning both; shoveled snow; mowed lawns; raked leaves; pulled weeds; and played a lot of sports.

In the 4th grade, his class had a new teacher who, on the first day of school, posed the question "What do you want to be when you grow up?" and asked each of the youngsters to rise in turn and divulge their future hopes. The girls aspired to be nurses, secretaries, or grade-school teachers. (How times have changed!) And the boys wanted to be policemen, firemen, or professional football players (easy to understand – all had been taught that "the policeman is your friend," and Columbus, after all, is the home of the Ohio State Buckeyes).

It came Billy's turn. He rose and with little hesitation announced: "I want to be a professor." Of the 25 students, he was the odd one out – he didn't "compute." But he was honest. (He's always found it curious that he said "professor," not "scientist." But he now understands – even at that early age, he regarded being a professor as a noble calling, a "job" that would actually pay money to a person to help educate others and thereby improve the society, while at the same time providing an opportunity to make new discoveries about the real world. Young Billy could never have said "scientist," a revered moniker far above his ken.)

2.2 BILL'S "BREAK-IN" TO THE SCHOOL GYMNASIUM

At about that same time, Billy started playing sports: basketball, baseball, football. In the winter, the snows came, and his Clinton School basketball team couldn't practice on weekends at the outside court of a nearby church. "Ahh," he thought, "the school gym is not being used." So, after school one late autumn Friday afternoon, he stood on a toilet in the boy's restroom in the basement, shimmied up the stall, and unlatched the window above. The next morning, with his basketball in hand, he lowered himself through the small window and opened a door to the school from the inside. Five or six of his team-members came through and they practiced.

The following week there were perhaps eight or ten of them. The two rules Billy made up were strict and never broken: (1) stay in the gym – do not enter any other room; and (2) if the cops come, disperse, fly like the wind, get out of here immediately!

All went well, until one of Billy's grade-school pals (he never knew which one) "mouthed off," spread the word. Within a couple of weeks, there were 30 or 40 kids in the gym, boys and girls both, all having uproariously great fun. For many weeks, everything continued to go well as these "self-policed" unsupervised little grade-schoolers followed the rules – no problems, no injuries, no fist-fights, no nothing!

Things remained that way until one Saturday morning when a teacher came by and looked in. Everyone panicked – the kids cleared out immediately scrambling pell-mell to the exits, and never tried it again. (None of them ever got trouble – perhaps the teacher thought this to be school-sanctioned event – but a year later, it led to the founding of the "CBA," the Clintonville Boy's Association, which provided venues and coaches for weekend neighborhood youth sports.)

In truth, James William Schopf, young Billy, never particularly liked being called "Bill," in those days it being rather common parlance to speak of "Just Plain Bill" (from the title of a long-running popular daytime radio drama). So, at the beginning of the 5th grade, he asked his teacher and classmates to call him "James" (he would have preferred Jim but that was how his father was known). Then, about half-way through the school year, the girls in his class discovered how to use a telephone. The calls started ringing-in at his home. And when the girlish little voice at the other end asked politely "to speak with James," his father repeatedly took the phone and said: "This is James. What can I do for you?" Billy's father, James Morton, thought this was terrifically humorous! To young Billy, however, it was enormously embarrassing.

On the first day of school in the 6th grade, he announced to the class that from then on, he should please be called "Bill."

2.3 JUNIOR HIGH

Instead of continuing on in public school for his junior high education, Bill's parents then enrolled him at University School, an experimental school for 7th–12th-graders run for the so-called "gifted" by Ohio State University's Department of Education. (Bill had no understanding of why he could not continue on with his Clinton School pals – though, perhaps, it may have had something to do with his IQ test scores.) At University School, he found some new girlfriends; his trumpet playing markedly improved; and he learned basic mathematics, photography, and even how to prepare "Baked Alaska."

But University School was odd – there were many very gifted kids but a sizeable contingent (5%–8%) of misfits, troubled youngsters placed there by their wealthy parents after their kids has been expelled from the city-run public schools. And as part of the experimental curriculum, the students voted on what subject matter they wanted to study (the idea being that if the students, not the teachers, decided the subject matter, they would be more committed to learning). Bad idea. Bill's Life Science class voted to learn about "deep-sea life," during which he learned hardly anything of lasting value – except that quite a number of abyssal animals are bioluminescent – whereas he missed out on the fundamentals of biology, which would later have been helpful.

Still, though Bill liked the school, his father did not. It turned out that University School did not give grades. Rather, the teachers wrote multi-page essays about each student, their strengths, weaknesses, abilities, deficiencies. Another bad idea! Reading all this stuff, which took the time of his father, who in any case was not much interested in it, annoyed Bill's dad. As his father put it, "How do I tell whether this is an A or an F?"

At the beginning of the 9th grade, after two years at this experimental school, Bill's parents transferred him back to be with his grade-school pals at Crestview Junior High. He did well reunited with his earlier friends; learned algebra (the only part of the math curriculum that he has ever really ever used); continued to improve his trumpet playing; and made "All City Honorable Mention" as a basketball player. This transfer back to a large public school was good for him.

Bill continued to carry newspapers, both morning (the *Ohio State Journal*) and afternoon (the *Columbus Citizen*) – both now defunct. During the summers, he played baseball (making the "All City" and "All District" teams) and spent a lot of time at the neighborhood public swimming pool (also now defunct), where he was on the swim team. He was pretty decent in the 50 yd (then yards, not meters!) breast stroke and 100 yd freestyle (though in the "IM," the individual medley, his backstroke never measured up). He taught swimming lessons and was a life guard (once even saving a comatose swimmer at the bottom of the diving pool).

The folks in charge of the public pool were terrifically kind and supportive. In fact, they gave him keys to the facilities. (Note that this would not happen now. He was only 14 years old, "too young" to be given such leeway, and had he misused this opportunity or become injured, the liability of the pool's owners would have been staggering.) So, every morning after finishing his paper route, he headed up to the pool where, alone, he swam back-and-forth across the pool for one, two, or three miles – half with a rubber tire inner-tube wrapped about his legs to improve his "pull," the other half propelled only by his legs, his arms outstretched and immobilized as he grasped the foam "kickboard" he pushed ahead to improve his "kick."

Near the end of one summer, Bill was taken to the "Nationals," that year held at Dayton, Ohio. He had no idea what the Nationals were – no one had bothered to tell him, it being to him just another swim meet but with older guys (in those days, he delivered the newspapers but didn't read even the sports section!) As a 14-year-old, he came in 5th, 1.2 seconds off the world record. He discovered later that the top three went on to the Olympics.

After the swim meet, he was approached by Mike Pepe, the swimming coach at Ohio State University (OSU), who told him that Bill was welcome to use the OSU pool to train with his team. To Bill that would have been fun – none of the Columbus high schools had such facilities, and he could have learned a lot from the older guys. But as he thought this over for the following two or three weeks, he came to realize that this was not for him (even given Pepe's inducement that "there might be a scholarship in it"). Though Bill had done well in the Nationals, he simply did not have the physical attributes of the other swimmers – tall bodies, long legs, large feet, long arms, large hands. Maybe he would grow, maybe not (as it turned out, he did not appreciably change). Anyway, Bill's plate was already full of other activities that at the time seemed to him more important. (Sadly, however, being an immature youngster, he never formally replied to Coach Pepe – now, much to his chagrin – and he doesn't recall that he ever mentioned this episode to his parents or anyone else). *Ahh ... throughout his early life, Bill was a lucky lad!*

2.4 COLUMBUS (OHIO) NORTH HIGH SCHOOL

After junior high, Bill then had (as he later came to appreciate) the privilege of moving on to Columbus North High School. This was a large school – 3,000 students in three grades (roughly 1,000 students more than the four-year college he later attended). It was also racially and socioeconomically diverse and the academically best high school in the city. He enjoyed high school.

At North High, thanks to Bill's band director, Mr. Les Susi, his trumpet playing continued to improve. Over earlier years (from the 7th grade on), he had participated in regional, district, and state-wide solo contests, consistently receiving a top-score "Superior" ranking. He played in the All Ohio Boys Band, touring around the Ohio State Fair, and as a high-school senior was selected to be First Chair in the All-Ohio Symphony Orchestra.

Despite the high academic ranking of Columbus North High, Bill did not learn a lot of lasting value. His chemistry teacher was a shambles (evidently more interested in recruiting Bill for the football team than teaching his subject). In physics, he learned that "F = Ma" (force = mass times acceleration), but that was about it. From his European history class, he can now recall only that the French–English Battle of Hastings occurred in 1066 and that the Magna Carta dates from 1215. Maybe he was not paying much attention! OK, his grades were good. But by that time he was an open bucket, ready to learn and actually interested such stuff, whereas many of his classmates (at this stage of life being "Bags of Hormones") evidently were not. Given the circumstances, the teachers were doing the best they could.

In fact, as Bill now looks back at those years, there were only two high-school courses that ever really helped him. One was Latin – no longer taught in most high schools, but for him continually useful as he figures out English cognates of Latin-derived scientific terms (e.g., arthritis – one of his current occasional maladies – derived from Greek *arthon*, "joint," via Latin, as in the biological term "arthropod," Latin, "jointed foot.")

You may find it odd, but the other high-school course that Bill values was typing. He was in a class of 45–44 girls with him being the sole male interloper. Bill was there because he knew that when he went on to college, he would be expected to write term papers. The girls were there because they wanted to have the skills needed to be secretaries (this, folks, was 1958). He did learn to type, reasonably well, which for him has been a big help ... as evidenced by the preparation of this book. But think for a moment about his situation. There he was ... one boy, 44 girls. For him, it should have been Nirvana. *Not so!* Three or four of his girlfriends were also in the class and he spent the entire semester worrying that they would find out about each other! (*Ha!*)

When Bill was a high-school sophomore, he wanted to buy a pink shirt so that he could fit in with the other guys in his class (this, coupled with a charcoal sports coat, being the current high-school style). He asked his parents: "Nope. No way. That's simply a passing fad. If you want it, go earn the money and buy it yourself."

This struck Bill as a bit unfair 'cause he was already earning money, still carrying the morning newspaper ... but those monies went to his "college fund" (which ultimately paid for his first two years in college). So be it. The next day, he ventured over to a local plumbing establishment and landed a job as their floor cleaner. After school, twice a week, he scrubbed the floors, stripped the wax, and resealed the floors. (And he bought his pink shirt!)

Bill then landed a "biggie job" – station manager for the *Columbus Citizen* newspaper, during the school year working 30 hours a week (upped later to 60 during the summer), overseeing the distribution of 2,200 papers by 24 paper carriers. After participating in a city-wide contest among the *Citizen*'s outlets to increase subscribers, in which his charges had come in second – his newsboys, spurred in part by the camaraderie-promoting "pizza parties" Bill provided for the group – went on to win five consecutive city-wide titles and by doing so earned trips to various "far-off" cities, most notably New York and Washington, D.C. (Bill's first ventures to these major hubs).

When Bill started this job he was only 16 – so, at least initially, by state law, he was permitted to work only six days a week. In the next year, at age 17, it was seven days a week, by which time he was leader of his North High dance band. Because the band was non-union (though by that time Bill was a dues-paying member of the Musicians Guild, mostly because he also played in various operas and with the Columbus Symphony), the band got lots of "gigs" at Ohio State fraternity parties. There were many Saturday nights when the band finished at 1 or 2am and Bill then – as the other band members went home to sleep – trekked down to his newspaper station to prepare the Sunday morning newspapers. To him, this was just fine – he had the old family car (a '49 Chevy, with a broken clutch) and was doing something useful, which his parents seemed to accept.

At the end of his junior year at North High, Bill was selected by the high-school faculty to be one of the two in his class to attend "Buckeye Boys State," a primer on American democracy and the US political system. He learned a lot and it was fun! At the Boy's State he was elected to the Supreme Court and naively decided that his real calling in life was to serve on the Supreme Court of the United States. Why? The answer was simple: Bill figured that a member of that court has only two responsibilities: first, know the law, thoroughly; and second, by interpreting the law fairly, help the society to improve.

Ultimately, of course, Bill did not follow the path to the judiciary – he could not have done so and he didn't even try. For him to have even been considered for such a position would have required him to "play politics," of which his father would never have approved. To Bill's father, the only truly honorable US president was Harry S. Truman and maybe or possibly or perhaps Franklin D. Roosevelt ("F.D.R.") – though in his father's opinion, the Roosevelt family had "far too much money" and was thus "suspect." Now, after reading a lengthy biography of Franklin Roosevelt (which Bill read shortly after the beginning of the 2008 "Great Recession" in order to fathom how the United States weathered the aftermath of the 1929 worldwide Great Depression), Bill is convinced that F.D.R. was first rate and that, except for his attempt to stack the Supreme Court with personally selected sycophants, he did a splendid job under the exceedingly difficult conditions of the recovery from the Great Depression and the onset of World War II.

To an extent, however, Bill still thinks that his father was right about politicians generally who, though commonly beginning their careers as well-intentioned idealists, too often devolve into self-serving opportunists who tell the populace "what they want to hear" not "what they need to know." Given our democratic system of government and the overall level of education of our citizenry, that ploy is easily understandable – perhaps even necessary for political survival – but it is easy to wish that things were different.

Because his older brother Tom had, at their dad's behest, had gone off to college a year early, for Bill's three years at North High he was the only son at home. But the house-rules did not change. From the 5th or 6th grade on, each evening at the beginning of dinner, Tom and he each had to report on "what you had learned today." When Tom was there, he always went first – giving Bill a chance to come up with something, anything, to say. (This nightly ritual was not much fun for either of the Schopf boys!)

Finally, in Bill's senior year, he thought that at long last he had this nightly problem nailed. His Advanced Calculus class had received grades from their mid-semester exam. Bill received the top grade of 97%, the second-highest in the class got 64%, and the others all failed. Even though Bill was miserable at math, he thought "Ha! I finally got 'em – my parents will be pleased." Not so! He said his piece. He sat. Utter silence. No response. Finally his father cleared his throat, paused again, and lowered the boom: "Where are the other three points?" His father expected perfection, and his second son had not measured up. Bill lost, not won! The few persons to whom Bill later recounted this tale all reacted negatively – "Your dad was incredibly mean," one even exclaiming "God! That's absolutely moronic!" But Bill immediately understood his father's lesson: "Good enough" is simply not good enough!

It soon became time to apply to colleges. Bill's list included three choices: Rice University in Houston (recommended by Dr. Gilbert H. Cady, a friend of his family and his father's mentor when he was at the Illinois Geological Survey); the University of Chicago (which appealed to Bill largely because its former president, Robert Maynard Hutchins, had established it as a first-rate academic institution); and Oberlin College (then the #1 small college in the country and where Bill's older brother, his hero, was already enrolled). The Ohio State University in Columbus was Bill's "back-up" in case he was not admitted by any of the others.

Bill was unaware that his high school had a counselor to help in the selection of colleges. Holy Smokes! He got plenty of "counseling" at home! But early in his senior year, he was asked to visit the counselor's office to meet a Harvard recruiter. The recruiter talked, Bill listened ... but Bill was skeptical. A few weeks later, the recruiter returned, evidently only to see Bill. The Harvard rep did his best "sell job," even telling Bill that he was "certain to be admitted, probably with some sort of scholarship." Bill did not commit, never applied, and in fact had no intention of doing so. He doesn't recall that he even told his parents about these interviews, because to his mind Harvard was a place for privileged "me-first" prep-school "Richies." Not Bill's style. He wouldn't fit. And he didn't want to be like them. (Later, when he was a Teaching Fellow as a Harvard graduate student and saw these students in action, he silently congratulated himself – for their first couple of years, these "best of the best" Harvard undergraduates are remarkably "full of themselves," maturing only in their later years.)

As it turned out, Bill was admitted to Rice, Chicago, Oberlin, and Ohio State. He first turned down Rice – which, when it was founded was tuition-free to all qualified applicants, as was Oberlin in its beginning days – primarily because he had misgivings about how a Mid-Westerner such as he would fit in the Texas milieu. Years later, the many months he spent in Houston during NASA's Apollo Program affirmed to him that Houston would not have been a good place for him to be. He might well have gone to Chicago – it had great appeal, despite being then plagued by the Blackstone Rangers, a notorious street gang. But the pull of Oberlin and the chance to overlap with his older brother were too strong. Bill selected Oberlin, for him a "perfect fit."

Near the end of his time at North High, the school faculty selected Bill as the "Outstanding Student" in his high-school class. This presumably was because of his trumpet playing, not his grades (Bill having ranked only 11th out of the several hundred in his graduating class). He never told any of his classmates about this accolade, not even his best girlfriend – but it pleased him because his father was required to attend the luncheon, along with the high-school principal, where Bill was honored. Fifty-five years later, Bill was the first in his class to be elected to the North High School Hall of Fame.

By the completion of high school, Bill was essentially independent and had come to know that imagination, diligence, and hard work, if it led to excellence, would be rewarded (his dad's expectation of perfection). The credit for this – which has carried Bill throughout his life – belongs to his parents, his older brother, and his teachers.

2.5 OBERLIN COLLEGE

Oberlin, a small (in Bill's day, only 2,000-student) liberal arts college in northeastern Ohio, is a remarkable institution. Like many small colleges founded in that part of the United States as its population moved westward, Oberlin has a strong religious heritage. The college and encircling community were founded in 1833 by a Presbyterian minister (the Rev. John J. Shipherd) and a missionary (Philo P. Stewart) to "train teachers and other Christian leaders for the boundless most desolate fields in the West." Guided by the ideas of the Alsatian pastor John Frederick Oberlin, who had pioneered educational programs and introduced the trades of masonry and blacksmithing throughout poor communities in France, the College adopted the motto "Learning and Labor." Tuition was free, and the students were expected to contribute by helping to build and sustain the fledgling community.

Over the years, Oberlin removed its formal religious affiliation, but it has never lost its underlying commitment to progressive causes and social justice.

The college was the first co-educational institution of higher education in the world (1833) and the first in the United States to be racially integrated (1835). In the 1850s, Oberlin was a major cog in the underground railroad that enabled escaping slaves to transit into Canada; in 1894, Oberlin was the founding home of the Anti-Saloon League, a major force in the Prohibition Movement that led to passage of the 18th Amendment to the US Constitution; and Oberlin College graduate Lucy Stone was a prominent figure in the fight for women's suffrage, leading groups such as the American Equal Rights Association and the American Woman Suffrage Association that, in 1920, led to the adoption of the 19th Amendment, which opened voting ranks to women.

By 1959, when Bill entered Oberlin, the college had a long history not only of societal activism but also of academic excellence. Some 52% of the members of his entering class, from all 50 states, were high-school valedictorians; and earlier in that decade, based on a national survey, the *Chicago Tribune* had named Oberlin the number one co-educational liberal arts college in the country, citing the college's exceptionally high standards of scholarship and teaching and its record of producing one of highest rates of graduates who go on to earn doctorate degrees. Even now, in 2018 – more than half a century later – Oberlin continues to produce more eventual Ph.D.s than any of its peer institutions, including at present some 23 members of the US National Academy of Sciences (NAS) (given Oberlin's small student body, a stunning one percent of the total NAS membership).

When Bill was a student, Oberlin's president was Robert Kenneth Carr, former president of the American Civil Liberties Union. As surprising as it may seem to someone not well acquainted with the college, even though Bill was "just another student," he came to know President Carr and his wife, Olive, both of whom were kind and generous to him, even inviting him to their home to share a luncheon with them and the commencement speaker at Bill's graduation. (Such openness to students was not all that unusual, Oberlin faculty on occasion also inviting undergraduates to dinner at their homes.)

President Carr was a quiet, thoughtful, unassuming intellectual – an excellent role-model for the student body – deeply committed to the betterment of Oberlin and of American society. And he had no intention of permitting the college to sit back, relax, and "rest on its laurels." To this day, Bill remember the title of an address he gave to the entire college community at Finney Chapel, the largest assembly hall on campus: "Nothing Fails like Success." Though President Carr was not a fiery orator, his address was on the mark, wholly correct. Oberlin had a century-long record of success. Great! But pressing problems still remained unsolved. He spurred the students on.

In the decades following Bill's attendance, Oberlin has continued to be at the forefront of significant social change: the integration movement of the 1960s; "women's liberation" in the 1970s and 1980s; and, more recently, liberalization of gender-based laws and practices. When Bill hitchhiked home from college ("safe," in his college days) the kind folks who picked him up would not uncommonly remark about "Oberlin Left-Wing Commies," a theme repeated by Nathan Hiller in a 2016 article in the *New Yorker* that described Oberlin as "a school whose norms may run a little to the left of [the avowed Socialist] Bernie Sanders."

It is no doubt true that some Oberlin students are intemperate and uncommonly imbued by an overriding sense of "political correctness." But it is also true – as is often said of Oberlin graduates – that "though they may not do well [financially] they are committed to doing good [for the society.]" OK by Bill. In his view, it is a great boon to our society for our idealistic youth to push, pull, and prod our country to improve. In a democratic society such as ours – one that continuously moves in fits and starts, three steps forward and two steps back – the seemingly small single steps forward ultimately add up. And throughout our history, the impetus for that forward progress has come from our impatient idealistic youth. Bill cheers them on!

Thanks to Bill's "college fund" – plumped markedly by his dance band "gigs" and his job as a station manager for the *Columbus Citizen* – sufficient funds had accumulated to pay tuition for his first two years of college ("big bucks," both then and now). During his first two years at Oberlin, he made spending money by carrying the college mail around the campus (by bicycle, no cars being permitted for Oberlin students – and no fraternities or sororities, either). In Bill's 3rd year, he was appointed a "Junior Counselor," responsible for mentoring 20, then 40 incoming freshmen (including as one of his charges Tom Liggett, who went on to become a member of the National Academy of Sciences and is currently a University of California Los Angeles [UCLA] Professor Emeritus and Distinguished Research Professor). And in Bill's 4th year, he was appointed to the then experimental position of "Student Master" of a newly built dormitory housing 200 upperclassmen – with both such jobs providing free room and board. All told, his college education did not cost his parents a huge sum of money.

2.6 BILL'S ONE-NIGHT GIG IN THE TOMMY DORSEY BAND

For most of Bill's time at Oberlin, he was the best trumpet player around (the Oberlin Conservatory, as excellent as it is and has always been, not being known for its Brass Department). Bill got together six or eight other guys and formed a band. Only a couple of them were Conservatory students, so as a group they had no right to practice in the Conservatory facilities. Nevertheless, as in his grade-school days, when he broke in to the school gymnasium so his basketball team could practice, Bill figured out a way to "beat the system." No one used the practice rooms during lunch. But if he and his pals skipped lunch they could, and they did! All went swimmingly for nearly two months – the band had a great time, enormous good fun – but what they had not foreseen is that unbeknownst to them, their better-and-better Dixieland Band began to draw outside crowds of students as the music filtered out.

The listeners first kept to the sidewalk on the other side of the street facing the Music Building. A couple of weeks, later the crowd increased and flooded the street. Someone called the cops. The Conservatory faculty complained, and the band was summarily kicked out, moving their practices to a local Episcopalian church that one of the band members attended. So it goes. Live and learn! (The irony of this is that the Oberlin Conservatory, nationally ranked among the top three with Julliard and Curtis, is now well known for its "Jazz Department." If this episode had any effect on the Conservatory, it was good for all concerned!)

Because Bill didn't have much competition from the Conservatory students, when the Tommy Dorsey Band came through to play for an all-campus dance and needed a stand-in for their second-chair trumpet (the "star" who plays all the stand-up solos), Bill was pegged for the job. He got big money (several hundred dollars); did what even then he thought to be a lousy ticky-tacky job on the solos (though the band members were kind to him); and learned about the life of a traveling musician, filled with narcotics, unfaithful spouses, strife, unhappiness, and boredom – except when the band was making music. Bill loved making music – the greatest "high" he had ever experienced – but such a life, even surrounded by such absolutely brilliant all-star musicians, was not for him.

2.7 FIRST YEAR SUCCESSES (AND A DISAPPOINTMENT)

During Bill's first year at Oberlin, he took a couple of Philosophy courses that he had always greatly valued, courses that, because they are founded on serious logical reasoning, he has frequently recommended to UCLA undergraduate science students. In the one of those courses, the professor assigned the students to construct and defend a simple syllogism. You will no doubt remember the format: "If *A* is the cause and the only cause of result *B*, and if result *B* does not occur, then cause *A* does not exist." A simple example: If the Sun is the cause of sunlight, and if sunlight does not occur, then the Sun does not exist.

The other students wrote about such themes as "a loaf of bread has gone missing from a grocery shelf; a man is running down the street with a loaf under his arm; the man must have stolen the bread." (Though, of course, the man may actually have paid for the bread and was simply trying to catch a bus!) In any case, Bill wasn't much interested in such matters, but he *was* interested in evolution. So, his syllogism: "If Darwinian Evolution is caused by, and only by, life's adaptation to a changing physical-biological environment, and if the physical-biological environment does not change, then evolution will not occur." In essence, this is the "null hypothesis" of Darwinian evolution, which must be true if it is a result only of adaptation to a changing physical-biological environment. The professor liked it, invited Bill to dinner at his home where Bill met his wife and children, and tried to entice Bill to major in Philosophy.

Even though Bill didn't follow up on the suggestion to major in Philosophy, he has always thought it was really, really kind of the Oberlin professor to have showered him with such attention. Yet, understandably, Bill could not have agreed to the professor's kind urgings. At best, Bill could tell that "good ideas" matter only if they are followed up by "good deeds." And Bill's hero, who tried to accomplish this feat, was the British philosopher Bertrand Russell (1872–1970), who had several times run for election to Parliament only to be repeatedly defeated. Nevertheless, Bill understands and deeply appreciates the value of such "good thoughts" – after all, it is just such notions, if they are put into practice, that will help societies throughout the world to become better and better.

Over his career, Bill has never forgotten the little syllogism he wrote for his freshman Philosophy class, but it's only recently that he found the evidence showing it to have backing. The basic problem was that to prove the point, one needs to find an unchanged ecosystem in a physical-biological environment that itself had remained unchanged over many hundreds of millions, even billions, of years. Earth's atmosphere won't work – in any given spot, it changes almost daily. The land surface cannot fit the bill – winds and rains erode it away and, due to the changing climate and the incessant movement of the continental masses, it changes markedly over time. And even the oceans won't work, with their omnipresent currents, daily fluctuations, and occasional tsunamis. The world we know changes – sometimes day to day, sometimes over a few hundreds or thousands of years.

Finally Bill found what he had imagined. In 2015, he described two examples of a previously completely unknown fossil ecosystem (sulfur-cycling sulfuretums), 1,800 and 2,300 million years old, in originally far off-shore deep-sea rocks, sediments preserved in the zone beneath the surface of submarine mud and a stable unchanging environment – where there is no churning by water currents, no light, no day–night cycle, no oxygen (subsurface mud, presumed to be a vacuous "dead zone," that had been ignored by others). And he was then able to compare these finds with their modern analogues (reported in 2007 from deep-sea mud off the coast of Chile) and to show that their complete fossilized multi-member biota had remained unchanged in this unchanging environment over a period of two billion years.

This discovery of a complete many-member biota unchanged over billions of years, inhabiting a static environment devoid of the influence of currents, day–night, or seasonal perturbations – the first test and promising evidence of the correctness of evolution's "null hypothesis" 150 years after Darwin presented his theory – received immediate worldwide media coverage (in the first 10 days after its publication being featured by some 300 news outlets across the globe).

In the second semester of Bill's Oberlin freshman year, he took his first full-fledged Biology course, in which his older brother, by now in his senior year, was also enrolled. Bill received an "A" on the mid-term exam and (no doubt triumphantly) strolled over to Tom's dorm room to tell Tom his results and find out how he had done. Tom received an "F" (a result, one might imagine, of "senioritis"). To Bill, this was a terrible disappointment! He bawled Tom out. And then Bill cried. Tom was Bill's hero. He had let Bill down! Tom soon got to work, received an "A" on the final exam and a "C" in the course. Bill continued to do well and then, at the end of the course, he was invited by the professor, Ed Kormondy, to be the professor's field assistant that summer doing limnology. *Ahh ... Bill once again lucked out!*

When Bill signed on to work with Professor Kormondy, he was not at all certain what "limnology" actually was. Nope, rather than involving studying the limbs of lithesome young beauties at the beach (as Bill might have hoped), it turned to be the biology-ecology of inland ponds (much like a little study he had done on his own during the summer between his junior and senior years of high school). Professor Kormondy's project was intriguing, investigating the molting stages of a dragonfly (*Tetraganeuria cyanosura*) that ranges from northern Michigan, where the nymphs go through their 12 molting stages (termed "instars") over a period of two years, to South Carolina, where they mature to adults in only one year. The question was, "Why the difference?" Day-length? Temperature? Different genetic strains? Improperly identified species?

This was interesting stuff. Bill learned the rudiments of freshwater biology – but he also learned a whole lot more. Toward the end of that summer, 1960, they journeyed to the Carolinas, and at gas stations, Bill was astounded to see separate drinking fountains for "whites" and "negroes," and three restrooms: "women," "men," and "negroes." He'd never heard of such a thing! It was a real eye-opener!

2.8 FORMATIVE YEARS ARE INDEED "FORMATIVE"

Over the (many) years that have followed these episodes, Bill has come to realize that such formative years – especially the grade-school years – are prime determinants of a person's development. It's far afield from his areas of expertise, but he has seen the data and observed the development of his schoolmates. Patterns of human behavior – including imagination, courage, and initiative – traits in humans are set early, just as they seem to be in other primates (e.g., chimpanzees).

Interestingly, this notion of behavioral patterns being set in place early in life harks back to the writings of the English philosopher and physician John Locke (1632–1704) – one of the most influential thinkers of the Age of Enlightenment and among the earliest British empiricists – who referred to it by the Latin phrase *tabula rasa*, meaning "blank slate." Locke's concept was that we humans are born without innate ideas or rules for processing our observations of and interactions with the surrounding world, and that such understanding is instead derived from gradually accumulated personal experience. Like Locke, Bill imagines that in the "nature vs. nurture" debate, the "nurture" side is determinant in the development of human values and social and emotional behavior – most notably, he thinks, between the ages of 3 and 8. But unlike some strict empiricists, he also credits the "nature" side of the equation – that is, human genetics – simply because we now know much more about the genetic bases of human behavior and that helps us to understand why it is that for some among us, the acquisition of new ideas and new knowledge seems to happen faster and be far easier than it is for others.

Importantly, of course, none of us get a vote in such matters – none of us gets to choose our parents, our genetics, our family setting, or our early-life experiences. Moreover, each of us is unique, one-of-a-kind. No one in human history has ever existed who is precisely like any single one of us, and unless we are cloned or have an identical twin, no one ever will again (at least for the "nature," gene-determined aspects of this two-part conundrum).

An obvious corollary follows: In human development, home/family life is paramount, and an early stimulating education is a close second. Lessons learned then are likely to have lasting value. A key to ultimate success, Bill thinks, is at early age to do something, anything, that pleases the adults, the "guiding light" mentors, and he doesn't think that it matters much just what – schoolwork, sports, music, writing, oratory, paid jobs, whatever. Each such task teaches the same simple lesson: You can only succeed if you keep trying to improve, time after time, and have the perseverance not to give up. For him, at least – in retrospect – he thinks this to have been a winning strategy.

From Bill's perspective, his "break-in" to the school gymnasium – as illegal, inappropriate, and wrong as it of course was – showed him, like his Iowa Chicken

Chase, that he was a problem-solver. At the time he never thought about any of that, but he now knows that those traits, courage, imagination, and a notion that "good enough" is not good enough have carried him though his life in academia and science. Those traits date from his upbringing and have served him well.

Though as a 1st-year college student, Bill had little experience and not much knowledge, his mind was open, a more-or-less empty void waiting to be filled, an ignorance that he has come to regard as surprisingly and importantly beneficial. It has pleased him, for example, to have had a notion about evolution's "null hypothesis" when he was a college freshman and then, 55 years later, to have found evidence that seems to fit – much less to have remembered to be on the lookout for such data over all those years!

From this, Bill has come to think that he and his fellow academics permit their 1st- and 2nd-year students to sell themselves short. The ignorance (*not* "stupidity") of youth can actually be a plus, not a minus, simply because young students have not yet learned the "accepted dogma." In fact, Bill has come to believe that if some "accepted truth" does not make sense to a thoughtful beginning undergraduate, it is likely to be mistaken (unless, of course, the professor has done a lousy job of explaining and justifying the claim). Breakthroughs in science are founded on curiosity, which necessarily requires the questioning of accepted wisdom – a cardinal trait of the youthful uninitiated because they do not yet "know" any better!

And because of Bill's ignorance, his trip to the Carolinas was both thought-provoking and deeply dismaying – he had never before observed societally accepted (much less institutionalized) racial segregation. He was, of course, aware of such tribalism as well as its close cousins, xenophobia and anti-this-or-that-religion (in his experience, principally anti-Catholic, anti-Jewish, anti-Mormon, anti-Muslim), but he had simply never seen such tribalistic racial segregation practiced and so obviously approved. To Bill, that was a rude awakening. Though our society has since made considerable progress, we all know that we have a ways to go; we're not there yet. Yet the lesson is clear: In this "three steps forward, two steps back" democracy of ours, the small single steps forward can, and do, ultimately add up.

3 Earth's Early Life – The Quest Begins

3.1 SECOND YEAR LIFE-CHANGER: DARWIN'S DILEMMA

At Oberlin, students were not permitted to "declare a major" before the halfway-point of their 2nd year, and though Bill was primarily interested in Biology – especially Evolutionary Biology – he opted to major in Geology. If it was good enough for his older brother, it would be good enough for him!

For Bill, the Oberlin system fit well. In fact, as an undergraduate, he actually took more course-hours in Biology than in Geology, his major, a background sufficient to enable him to be admitted to the Harvard Biology Department for graduate studies (where he took more hours in Geology than Biology, augmented by courses in organic chemistry and geochemistry). Bill's Oberlin liberal arts education served him well.

When Bill was a second-year college student enrolled in the introductory Historical Geology class, his professor, Larry DeMott, noted almost in passing that the Precambrian earliest history of life was completely unknown – and it had been since Darwin first raised the problem more than 100 years earlier. DeMott opined that "This is the greatest unsolved problem in all of Natural Science."

This was news to Bill! The oldest fossils then known were Cambrian-age trilobites, "lobster-like" arthropods far too complicated to have anything whatever to do with the earliest forms of life. But evolution was a fact, not some "Fake News" fable, so the "missing" Precambrian, pre-trilobite record of life just didn't make any sense. Bill told himself that this was something that he had better find out about.

After class, Bill returned to his dormitory room, took from the shelf his paperback edition of Darwin's 1859 *On the Origin of Species* (John Murray, London), and found that Darwin had addressed the issue, repeatedly, but could offer no solution. In Darwin's words, "There is another and allied difficulty, which is much more serious. I allude to the manner in which species bellowing to several of the main divisions of the animal kingdom suddenly appear in the lowest known fossiliferous rocks. … Consequently, if the theory [of evolution] be true, it is indisputable that before the lowest Cambrian stratum was deposited long periods elapsed, as long as, or probably far longer than, the whole interval from the Cambrian age to the present day; and that during these vast periods the world swarmed with living creatures. Here we encounter a formidable objection; for it seems doubtful whether the earth, in a fit state for the habitation for living creatures has lasted long enough. … To the question why we do not find rich fossiliferous deposits belonging to these assumed earliest periods prior to the Cambrian system, I can give no satisfactory answer … the difficulty of assigning any good reason for the absence of vast piles of strata rich in fossils beneath the Cambrian system is very great … The case at present must remain inexplicable; and may be truly urged as a valid argument against the views here entertained" (Chapter X, "On the Imperfection of the Fossil Record").

The next afternoon, Bill hiked over to the Oberlin Library – at the time, the largest small college library in the United States – and he began to read everything he could find on the subject. He soon learned about the acrimonious century-long history of the problem, but given the scant number of publications on the subject, his little "library research project" was by no means overwhelming. Over the following weeks, Bill discovered that there had been numerous possible-maybe-perhaps "discoveries" and well-intended hopeful claims, all unsubstantiated. But though there were few solid facts and hardly any useful articles, he also discovered that there were only two then-living American scientists who had written about aspects of the problem (one article each): Preston Cloud, then chair of the Department of Geology at the University of Minnesota; and Elso Barghoorn, professor of Paleobotany at Harvard. Bill wrote a letter to each expressing his interest in the problem. Both graciously responded.

To determine whether he might be admitted to do graduate work, Bill consulted university catalogs. Admission requirements at the University of Minnesota were based largely on course grades. He figured that he would probably be OK. But Harvard presented a problem since they only admitted "Honors Students." Bill didn't know what that meant – to him it was a new term. He looked it up in the Oberlin catalog and found that an Honors Program designed to encourage undergraduate research that would conclude with the production of an "Honors Thesis" had been established in the Geology Department in 1908, nearly half a century earlier. He then approached the chair of the department, Professor Fred Foreman, and was told that "This is a program we have never used and never will – undergraduates have absolutely no business doing research." Nevertheless, and despite Professor Forman's reticence, after Bill explained his plight, the professor was kind enough to suggest that Bill prepare a letter to the department faculty outlining his proposed Honors project. He did so.

At the time, the Oberlin Geology Department had three faculty members: mineralogist-economic geologist Fred Foreman (chair); geochemist-geochronologist Jim Powell (a Pat Hurley-MIT Ph.D. who was interested in Bill's work); and paleontologist Tony Gordon (filling in on a one-year stint for Larry DeMott, who had departed to establish a Geology Department at Knox College in Galesburg, Illinois). The faculty voted on Bill's proposal. In a split decision (one pro, one con, one abstention), Bill was permitted to be the Geology Department's first "Honors Student" (a now highly effective program, as Bill later discovered when he served as an Oberlin Trustee, that currently produces six to ten Honors Students per year).

Under normal circumstances, Paleontology professor Gordon would have advised Bill's Honors work – but he was between permanent jobs, having come to Oberlin as a short-term fill-in from the University of Puerto Rico (where, he told Bill, students in his classes had wrapped an arm around his shoulders and called him "buddy" – a familiarity he detested). He wanted to have nothing whatever to do with undergraduate students, including Bill. He was honest. Bill respected that.

OK. Sink or swim, Bill was on his own. From Professor Barghoorn, he obtained a chunk of the 1,900-million-year-old Gunflint chert (the source of Barghoorn's one

1954 co-authored paper in the field); cut the rock and ground thin sections; did the microscopy and found diverse fossil microorganisms (some new to science, some still not published); taught himself photomicrography; and prepared his Honors Thesis.

3.2 WELL-MEANING DISSUADERS OF BILL'S QUEST

During Bill's college years, he was repeatedly urged by well-meaning professionals, chiefly his father's friends, to cease or at least delay his efforts to uncover life's early "missing" fossil record. Bill's goal didn't fit "accepted dogma," and given that there had been no meaningful progress for a full century, it was feared that he was certain to fail – he would be wasting his potential.

Among numerous such episodes, one stands out. In April of Bill's sophomore year, when he was 18 years old, he was home for Easter recess. His dad was visited by Bill Chaloner (then a fairly recently minted Ph.D. who went on to become a Fellow in the Royal Society and an internationally renowned paleobotanist). Chaloner, Bill's dad, his brother, and he went on a field trip to collect plant fossils. Rain set in. Bill's brother and his dad huddled under one poncho and Chaloner and Bill under another. Chaloner was kind to Bill, and when they ran out of the normal pleasantries, Chaloner asked Bill what he wanted to do in science. Bill told him. Chaloner's advice: "Finish college [2½ to go]; get a Master's degree [2 years] then a Ph.D. [4 years more]; land a job as an Assistant Professor; and then [6 years later] as a tenured Associate Professor – when you can't be fired if you fail – work the problem." Silently, Bill totaled up the delay – 14 or 15 years – almost the length of his entire life! Bill thanked Dr. Chaloner for his kindness and advice (which, as a head-strong know-nothing 2nd-year college student, Bill had no intention of following). *Yes! Yet once again Bill was lucky!*

3.3 THE ROOTS OF DARWIN'S THEORY

As Bill became increasingly committed to discovering then unknown Precambrian record of life – thanks primarily to his Oberlin Honors research project – he became increasingly interested in how and why it was that Darwin created his theory of evolution. Like Bill's high-school teachers and Oberlin professors (and virtually everyone else with whom he had come in contact), he assumed that it originated *de novo* during Darwin's world-touring trek as a naturalist on the *H.M.S. Beagle*. As Bill learned during his continuing visits to the Oberlin library, this is not exactly true.

In fact, for Charles Robert Darwin (1809–1882; Figure 3.1), the inklings of his theory date back at least to the insights of his grandfather, Erasmus Darwin (1731–1802; Figure 3.2) – who died seven years before Darwin was born – an English physician who in 1796 authored *Zoonomia: The Laws of Organic Life* (Figure 3.3), which outlines the basics of evolution. And only a few years later, in 1809, Jean-Baptiste-Pierre-Antoine Lamark (1744–1829; Figure 3.4), a naturalist at the Muséum National d'Histoire Naturelle in Paris, published *Philosophie Zoologique* (Figure 3.5), which presented the first comprehensive theory explaining life's evolutionary progression. Not surprisingly, given the European belief-systems of the early 1800s, by then dominated for nearly two millennia by Christian Roman Catholicism, Lamark was

FIGURE 3.1 Charles Robert Darwin.

FIGURE 3.2 Erasmus Darwin, Charles Darwin's grandfather.

ZOONOMIA;

OR,

THE LAWS

OF

ORGANIC LIFE.

VOL. II.

By ERASMUS DARWIN, M.D. F.R.S.

LONDON.

FIGURE 3.3 Title page of Erasmus Darwin's *Zoonomia.*

FIGURE 3.4 Jean-Baptiste Lamark.

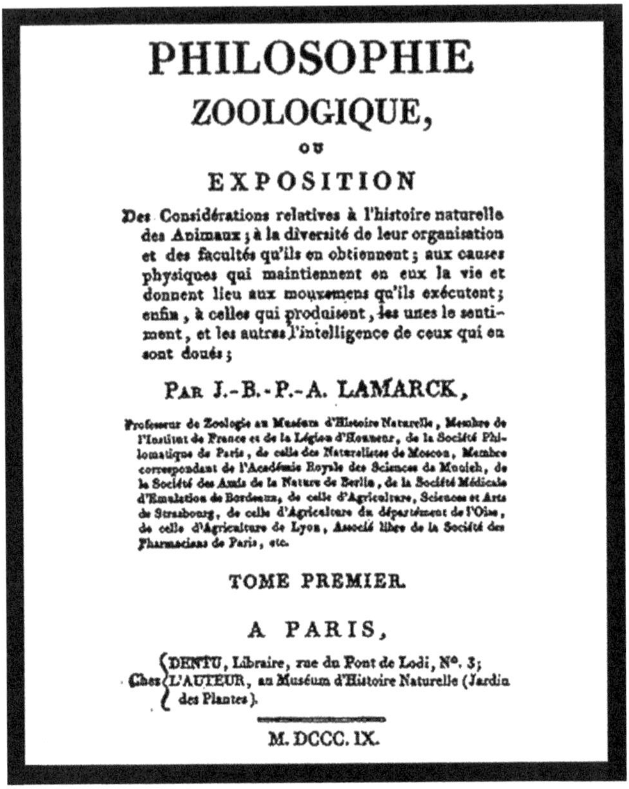

PHILOSOPHIE
ZOOLOGIQUE,
OU
EXPOSITION

Des Considérations relatives à l'histoire naturelle des Animaux ; à la diversité de leur organisation et des facultés qu'ils en obtiennent ; aux causes physiques qui maintiennent en eux la vie et donnent lieu aux mouvemens qu'ils exécutent ; enfin, à celles qui produisent, les unes le sentiment, et les autres l'intelligence de ceux qui en sont doués ;

PAR J.-B.-P.-A. LAMARCK,

Professeur de Zoologie au Muséum d'Histoire Naturelle, Membre de l'Institut de France et de la Légion d'Honneur, de la Société Philomatique de Paris, de celle des Naturalistes de Moscou, Membre correspondant de l'Académie Royale des Sciences de Munich, de la Société des Amis de la Nature de Berlin, de la Société Médicale d'Emulation de Bordeaux, de celle d'Agriculture, Sciences et Arts de Strasbourg, de celle d'Agriculture du département de l'Oise, de celle d'Agriculture de Lyon, Associé libre de la Société des Pharmaciens de Paris, etc.

TOME PREMIER.

A PARIS,

Chez DENTU, Libraire, rue du Pont de Lodi, N°. 3 ;
L'AUTEUR, au Muséum d'Histoire Naturelle (Jardin des Plantes).

M. DCCC. IX.

FIGURE 3.5 Title page of Lamark's *Philosophie Zoologique.*

repeatedly lampooned in the Parisian press. And he sometimes still is, as he was by the professor in Bill's freshman Oberlin Biology course, who derided Lamark because he attributed such evolutionary changes to the use/disuse of various traits that resulted in the transmission of acquired characteristics driven by some unexplained "vital force" toward perfection. (Bill later came to understand that the professor was off base. How could Lamark have done otherwise? He had no knowledge of the genetic bases of inheritance from generation to generation – not understood until 1900, nearly a century later, with the discovery of the notebooks describing the 1865 experiments on the heredity of pea plants carried out by Austrian Monk Gregor Mendel.)

Soon after its publication, Lamark's 1809 notion drew the wrath of Baron Georges Cuvier (1769–1832; Figure 3.6), a highly influential French naturalist, zoologist and professor at Collège de France, Jardin des Plantes, Paris. Among Cuvier's numerous well-deserved "claims to fame," he was Napoleon Bonaparte's Minister of Education and is widely regarded as a founder of modern Comparative Anatomy. Like virtually all of his academic contemporaries, Cuvier was a confirmed Christian Creationist.

But Cuvier had a problem. His studies of the fossils of the Paris Basin had convinced him that biologic species (Biblical "types") repeatedly originated, rose

FIGURE 3.6 Baron Georges Cuvier.

to success, and became extinct, a new idea at odds with accepted Biblical truth. His interpretation was reinforced by his comparative studies of the teeth and bones of a woolly mammoth from Siberia and the "Ohio Animal," a North American mastodon (a name, meaning "breast tooth," that Cuvier coined in 1817 for the nipple-like projections on the crowns of its molars) shipped to him by US President Thomas Jefferson (who is credited as the founder of North American Vertebrate Paleontology). Cuvier regarded these gigantic animals to be two similar but obviously distinct relatives of modern elephants that had become extinct. To fit such extinctions with then-prevailing views – the demise of Biblical "types" being at odds with God's plan for life on Earth – Cuvier proposed the new idea that the world had suffered through a repeated series of catastrophic extinctions followed by successive Biblical creations.

Cuvier's notions, like those of Lamark, were summarily pooh-poohed by Bill's Biology professor, mostly, he later came to understand, because the professor did not put himself in Cuvier's shoes, in his time and culture, and evaluate what was then "accepted dogma." Think about it. We, today, commonly accept as "truth" numerous concepts that are poorly or not at all established by observable facts (e.g., cosmological "dark matter," "dark energy," "string theory," and even, perhaps, the Big Bang-following "inflation theory"). One or two hundred years from now, humans will be making advances that we currently have no way now to fully understand.

The history of science, of ideas, is a great teacher, if it is evaluated objectively. Charles Darwin learned from the writings of his grandfather, Erasmus, and from Lamark and Cuvier as well. He benefited and incorporated their findings into his. The past and present are keys to the future – the Rosetta Stones of Life; if we learn from them, knowledge will advance.

3.4 DARWIN'S COLLEGE DAYS AND THE *H.M.S. BEAGLE* EXPEDITION

Both Darwin's father and his grandfather were noted well-to-do physicians, a profession representing, together with the clergy, the best-educated of early-1800s British society. Charles Darwin's fate was predetermined – he would become a physician. Thus, in 1825, he entered Edinburgh University in Scotland to study medicine. In the first class of this medical training, he is reputed to have witnessed a cadaver being cut-up to expose its innards – and he then immediately exited the lecture room and threw up. Whether this story is apocryphal or not, he soon withdrew from medical studies and in 1828 entered Christ's College, Cambridge University, England, evidently to become a minister – the only other highly educated profession of which his family (and particularly his father) would approve. Soon, however, he discovered that this, too, held no interest for him. Nevertheless, he continued his studies at Cambridge, attended Professor John Stevens Henslow's course in Botany, started assembling what was to become a magnificent collection of beetles, and read widely. In short, he became educated, receiving his Bachelor's degree in 1831.

Rather than becoming a physician or a cleric, Darwin had become enamored by the Natural Sciences, chiefly though the influence of the geologist-botanist Henslow (1796–1861) and Darwin's acquaintance with two other greats, first with Charles Lyell (1797–1875; Figure 3.7) – the famed "Present is the Key to the Past" geologist who, years later, was formally knighted – and subsequently with Joseph Dalton Hooker (1817–1911; Figure 3.8), one of the most accomplished British botanists

FIGURE 3.7 Sir Charles Lyell.

FIGURE 3.8 Joseph Dalton Hooker.

and explorers of the 19th century and arguably Darwin's closest friend. In August of 1831, Darwin, a 24-year-old recent college graduate, received news that a British Naval vessel, the *H.M.S. Beagle*, was preparing to sail from Plymouth at the southern edge of England for a round-the-world tour and was seeking an (unpaid) gentleman naturalist.

The lack of a salary was not a concern – Darwin's family had ample money. His problem was to obtain parental consent (not different from the problems 24-year-olds have today). He asked his father and received a firm "No!" But by this time, the voyage had become Darwin's dream. He turned to his uncle, Josiah Wedgwood II (son of the Wedgwood Pottery Works founder) and asked for his help. Josiah, the brother-in-law of Darwin's father (Robert Waring Darwin) intervened, and Darwin was permitted to sign on as the *Beagle*'s naturalist.

The voyage of the *H.M.S. Beagle* (Figure 3.9) lasted nearly five years (December, 1831 to October, 1836) rather than the two initially envisioned, and was intended primarily to survey the "New World" of South America, heading first to Brazil, then to Uruguay, around Cape Horn, up the western South American coast to Chile and the Galapágos Islands west of Ecuador, across the South Pacific to Australia, around Africa's Cape of Good Hope, and then finally heading home to Plymouth. On the return trip after circling the globe, the *Beagle* stopped again in Brazil to pick up raw materials to transport to England where they could be turned into commodities that were then sold in England and its numerous worldwide colonies, supporting its ongoing industrial revolution.

Throughout the voyage, Darwin and Captain Robert FitzRoy dined together nightly in the Captain's cabin, the only two on the craft regarded officially as

FIGURE 3.9 Model of the *H.M.S. Beagle.*

"Upper Class" and thus qualified for this privilege, a separation from the *Beagle*'s seamen being not only regarded as appropriate but also demanded by the established norm. Both Darwin and Captain (later Vice-Admiral) FitzRoy (or, perhaps more likely, the seamen under FitzRoy's command) actively gathered natural history specimens as the *Beagle* journeyed the world, the Captain having been so tasked as a representative of the Royal Navy. FitzRoy's collections of bird skins, nearly 200 specimens, were given to London's British Museum [Natural History] where virtually the entire collection still survives. (An interesting if little known fact about FitzRoy is that in 1865, six years after publication of the *Origin*, he committed suicide at his residence on Onslow Square, Kensington, a few blocks distant from London's Natural History Museum, after becoming distraught for having helped Darwin "disprove the Bible.")

Many attribute Darwinian Evolution to the insight he gained by his studies of "Darwin's Finches" on the Galapágos Islands, some 600 miles west of Ecuador. In part, at least, this is untrue. Rather, the records show that when the *Beagle* finally returned to England after its long voyage, Darwin delivered virtually all of his specimens to others to study. During the voyage, Darwin had collected nearly 500 bird skins as well as whole birds preserved in alcohol ("spirit"), various bird parts, and a few nests and eggs, the majority of which were turned over to the museum of London's Zoological Society (ZSL). There the collection was examined by the distinguished ornithologist John Gould (1804–1881; Figure 3.10), leading to the description of some three dozen new species and subspecies. In fact, it was evidently Gould, not Darwin, who recognized that the Galapágos finches were closely interrelated (Darwin having originally recorded them as being from a range of different taxonomic families). In 1855, the ZSL museum was dissolved and its collections were

FIGURE 3.10 Ornithologist John Gould.

dispersed, some 200 specimens being now archived at London's Natural History Museum. These and other of Darwin's specimens there archived are available for study by scientists and collections management specialists who can borrow them for research (but, like the world's oldest assemblage of microbial fossils Bill archived at the Museum in 1993, only if they are hand-carried from place to place, rather than being sent through the mail).

3.5 GENESIS OF THE *ORIGIN*

Darwin's diary records that in 1837, a couple of years after his return to England, he began writing his volume, and that on 3 October 1838, "for amusement," he read Thomas Robert Malthus' (1766–1834) book-length *Essay on the Principle of Population*, gaining insights that proved pivotal to helping him explain his observations. Darwin was already well acquainted with artificial selection – the domestic breeding of roosters, for example – and knowledgeable also about Cuvier's concept of extinction and the fossil record of animals, the broad outlines of which were well understood by the 1840s (Figure 3.11). But from reading Malthus' volume (Figure 3.12) Darwin grasped the unifying cause of these seemingly disparate observations, namely that life worldwide competes for the limited resources available, a competition that would necessarily result in a "natural selection" of the best adapted to the locally prevailing conditions – the winners would produce more offspring, and the losers would be culled from the population.

Darwin completed a preliminary draft of his volume in 1844, originally titled *Abstract of an Essay on the Origin of Species and Varieties by Natural Selection*

FIGURE 3.11 *The Antediluvian World* (James Reynolds, London, 1849).

FIGURE 3.12 Thomas Robert Malthus' *Essay on the Principle of Population.*

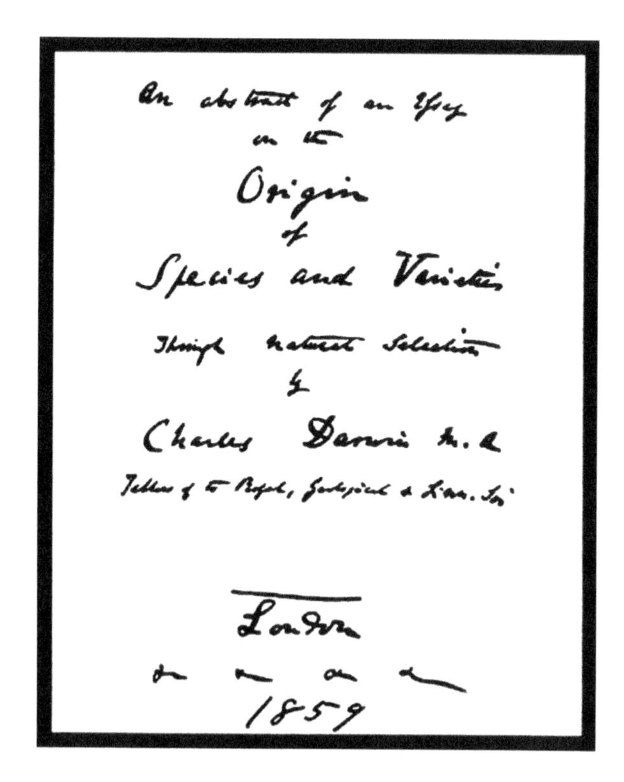

FIGURE 3.13 Title page of Darwin's original draft of *The Origin of Species.*

(Figure 3.13). But he delayed submitting it for publication for the following 15 years, allegedly to avoid offending his wife, a member of England's aristocratic (and devout Christian) Wedgwood family. Thus, it was not until the spring of 1859 that his world-shaking opus finally appeared – now bearing the revised extended title *On the Origin of Species by Means of Natural Selection, or the Preservation of Favoured Races in the Struggle for Life* (Figure 3.14).

Moreover, it seems clear that it would not have appeared even then, had he not received a letter in May or June 1858 from Alfred Russel Wallace (1823–1913; Figure 3.15), an English naturalist who, from his ongoing work in Malaysia, had independently arrived at the same conclusions about natural selection as had Darwin. Remarkably, like Darwin's theory, Wallace's ideas had been stimulated by his own reading of Malthus, evidently in about 1846, but he had not realized its import for explaining evolution until he lay recovering from fever in Malaysia a dozen years later. A principle component of Malthus' thesis was that if a population was held in check by limited resources, many of its members would not survive to reproduce. As Wallace wrote, "It suddenly flashed upon me … in every generation the inferior would inevitably be killed off and the superior would remain – that is, the fittest would survive."

Wallace's letter enclosed a short essay outlining his theory of natural selection, essentially identical to Darwin's. Darwin explained the situation to his geologist-botanist friend Hooker (by this time director of England's Royal Botanical Gardens

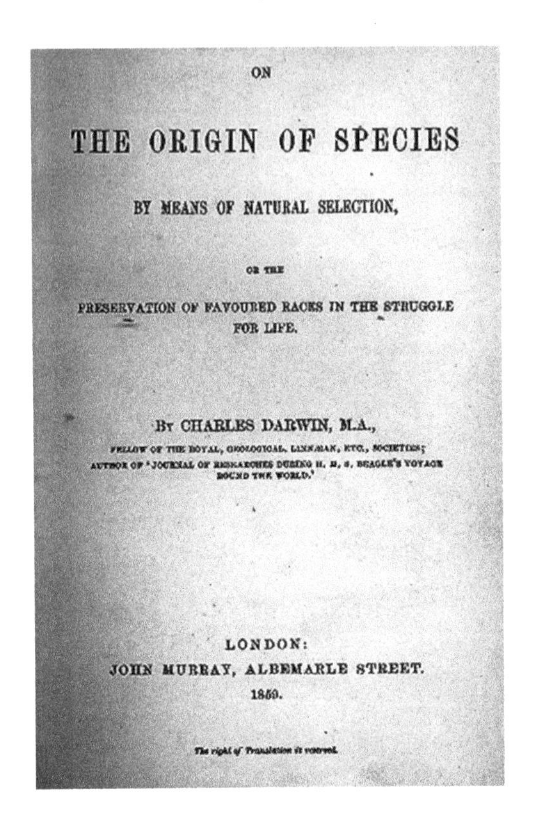

ON

THE ORIGIN OF SPECIES

BY MEANS OF NATURAL SELECTION,

OR THE

PRESERVATION OF FAVOURED RACES IN THE STRUGGLE
FOR LIFE.

BY CHARLES DARWIN, M.A.,

FELLOW OF THE ROYAL, GEOLOGICAL, LINNÆAN, ETC., SOCIETIES;
AUTHOR OF 'JOURNAL OF RESEARCHES DURING H. M. S. BEAGLE'S VOYAGE
ROUND THE WORLD.'

LONDON:
JOHN MURRAY, ALBEMARLE STREET.
1859.

The right of Translation is reserved.

FIGURE 3.14 Title page of the first edition of Darwin's *The Origin of Species.*

FIGURE 3.15 Alfred Russel Wallace.

FIGURE 3.16 Commemorative plaque, Burlington House, Linnean Society of London.

at Kew) and in mid-June 1858 wrote a letter to Lyell, the internationally renowned geologist, in which Darwin enclosed Wallace's essay and lamented his personal fate of having been scooped. Within a scant two weeks, Hooker and Lyell came to Darwin's rescue. Both having been privy to Darwin's theory for several years, they arranged for the Darwin and Wallace ideas to be presented back-to-back at the then almost-immediately upcoming July 1, 1858 meeting of the Linnean Society of London (Figure 3.16). (Interestingly, the Linnean Society, the oldest extant biological society in the world, is centered in Burlington House, a complex of buildings that is also the home of the Geological Society of London, for which Darwin served as secretary despite his lack of formal training in geology.)

Bill Schopf, as a now long-time Foreign Member of the Linnean Society – for which the Society requires that he append the honorific "F.M.L.S." to his correspondence with them – has been permitted to peruse their archives. From this, he has learned that the Lyell–Hooker submission letter reads in part: "These gentlemen [Darwin and Wallace] having, independently and unknown to one another, conceived the same very ingenious theory to account for the appearance and perpetuation of varieties and of specific forms on our planet, may both fairly claim the merit of being original thinkers in this important line of inquiry; but neither of them having published his views, though Mr. Darwin has for many years past been repeatedly urged by us to do so, and both authors having now unreservedly placed their papers in our hands, we think it would best promote the interests of science that a selection from them should be laid before the Linnean Society." Bill also learned – to his surprise – that these presentations elicited virtually no response from those assembled. The uproar was not to come until the publication of Darwin's epochal volume some nine months later.

The reason for here outlining the history of Darwin's seminal contribution to human knowledge – which historians of science have recounted repeatedly in appreciably more detail – is simply to illustrate a great truth about the workings of the scientific enterprise: Namely, that Kuhnian paradigm shifts – in this case from a Biblical- to a Nature-based development of life on Earth – do not arise out of "thin air."

Rather, much like the evolution of life itself, they build on that which has come before. Here, not only did Darwin evidently learn from the writings of his grandfather, Erasmus, and those of Lamark and Cuvier, but both he and Wallace found their defining inspiration in the insights of Malthus.

The bottom line: In science, this wonderful ongoing "detective game" we humans have designed, each of those who try to make meaningful contributions relies on workers who have come before. Darwin's ideas were not purely, entirely, original with him. And that is also true about those who have contributed to the paradigm shift resulting from the discovery of the solution to Darwin's "inexplicable" quandary – the so-called "missing" early record of life – the focus of this book. That is how human knowledge progresses. That is the way of science.

4 The Missing Precambrian Record of Life

4.1 A CENTURY OF GESTATION: 1859–1960s

Bill Schopf's quest to discover the "missing" record of Precambrian life was born nearly 60 years ago with his college Paleontology professor's assertion that "This is the greatest unsolved problem in all of Natural Science" and was spurred by Darwin who in *The Origin* stated his dilemma, namely that if his "theory be true, [the absence of a Precambrian fossil record] is inexplicable … a valid argument against [his] views." This absence was inexplicable to Bill, too, so he set out to find the answer. (Interestingly, various versions of this quote – revealing that Darwin, himself, had serious misgivings about his theory of evolution – were prevalent in the anti-evolution Creationist literature well into the 1970s, after which, due to the breakthrough findings of Precambrian fossils in the mid-1960s, they appear to have essentially vanished.)

For a full century, from the statement of Darwin's dilemma in 1859 to the game-changing finds of the mid-1960s, the more than three billion year–long Precambrian history of life remained unknown and was thought unknowable, the relevant facts unproven and unaccepted. As summarized below, this long history leading to the ultimate discovery of the solution to the problem featured a complicated series of starts and stops – well-intended culture-based errors; sound findings only to be later cast aside; authoritative but misguided "Science by Assertion"; and even international politics–based prejudice. The following are the highlights of the 100-year gestation of the development of this paradigm shift in the understanding of life's deep history.

4.2 FORTY YEARS OF TURMOIL: J.W. DAWSON AND *EOZOÖN CANADENSE*

The saga began in 1858, a year before the publication of Darwin's *Origin*, when a collector for the Geological Survey of Canada brought to Sir William Edmund Logan (1798–1875), the director of the Survey, some unusual rock specimens collected along the Ottawa River from the Precambrian Grand Calumet Limestone. The Grand Calumet was guessed (correctly) to be about a billion years old, some 500 million years older than the oldest fossils then known, and because Logan thought that these curiously layered specimens might be remnants of exceptionally ancient life, he displayed them at the 1859 meeting of the American Association for the Advancement of Science. In 1862, he brought them to England for examination by the ranking experts (a common practice for "big finds" in the colonies, England and particularly London being the intellectual hub of the worldwide British Empire).

Logan departed London empty-handed – the Brits seemed unimpressed – and little was done with these enigmatic specimens until two years later, when Logan observed similar forms in the Grenville Limestone near Ottawa, which, like the Grand Calumet, was about a billion years old.

That year, 1864, these new specimens along with the 1858 collection were delivered to Sir John William Dawson (1820–1899; Figure 4.1), the principal of McGill University in Montreal and, like Logan, a famous Canadian geologist. Indeed, and though Dawson was not the first principal of McGill, it is he who is credited with having raised the university to its long-held status as among the foremost such institutions in Canada. A devout Calvinist Christian of Scottish descent, Dawson studied the specimens and went further than Logan, identifying them to be giant foraminiferans (shelled single-celled protozoan animals), which in 1865 he formally named *Eozoön Canadense*, the "Dawn Animal of Canada" (Figure 4.2).

Eozoön entered the scene only a scant six years after publication of Darwin's *Origin*. Darwin's notions had raised a furor, and the very existence of such exceedingly ancient fossils caused Dawson angst. From his Calvinist teachings, he "knew" that Darwin's evolution was nonsense, yet he was confronted by this new find – fossil protozoans hundreds of times larger than those living today and vastly older than any of the various "types" mentioned in the Bible. Like Cuvier and his discovery of non-Biblical extinctions a half-century earlier, Dawson had a problem. How could *Eozoön* be made to fit with Biblical truth?

In his 1875 volume *The Dawn of Life, Being the History of the Oldest Known Fossil Remains, and Their Relations to Geological Time and to the Development of the Animal Kingdom* (Figure 4.3), Dawson states his case: "There is no link whatever to connect *Eozoön* with younger fossils … all stand before us as distinct creations.

FIGURE 4.1 Sir John William Dawson.

FIGURE 4.2 Dawson's *Eozoön Canadense.*

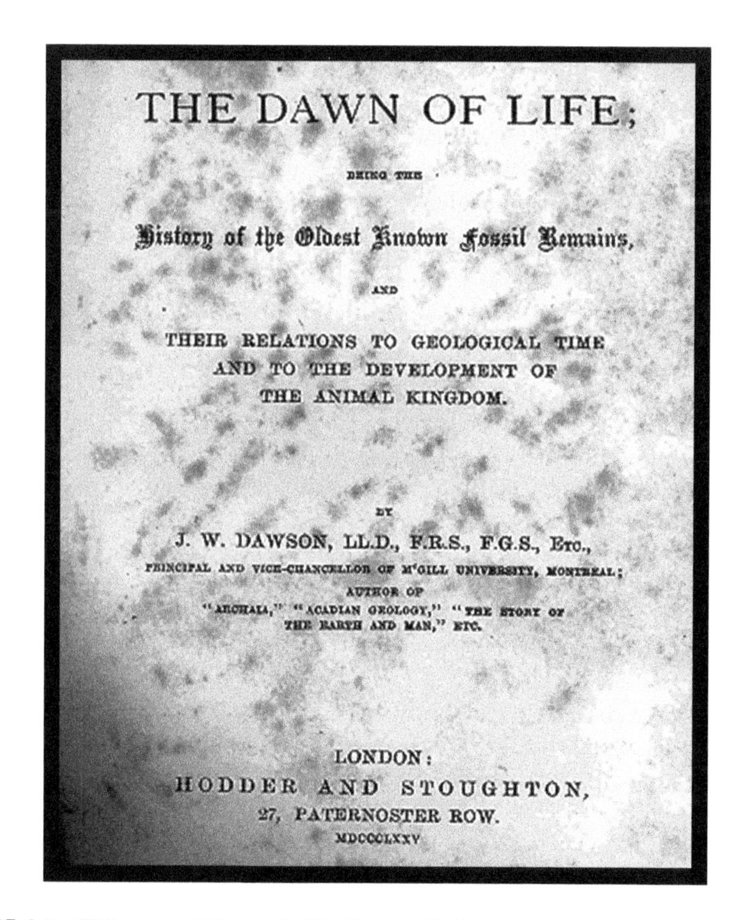

THE DAWN OF LIFE;

BEING THE

History of the Oldest Known Fossil Remains,

AND

THEIR RELATIONS TO GEOLOGICAL TIME
AND TO THE DEVELOPMENT OF
THE ANIMAL KINGDOM.

BY

J. W. DAWSON, LL.D., F.R.S., F.G.S., ETC.,
PRINCIPAL AND VICE-CHANCELLOR OF M'GILL UNIVERSITY, MONTREAL;
AUTHOR OF
"ARCHAIA," "ACADIAN GEOLOGY," "THE STORY OF
THE EARTH AND MAN," ETC.

LONDON:
HODDER AND STOUGHTON,
27, PATERNOSTER ROW.
MDCCCLXXV

FIGURE 4.3 Title page of Dawson's *The Dawn of Life.*

Eozoön thus bears damaging negative testimony against evolution, [a concept] incapable of proof and contrary to fact. Evolutionists are mere dreamers, having no scientific basis for their dogmas."

Turmoil ensued. The first challenge to Dawson's 1865 interpretation came from William King and Thomas Rowney, two mineralogists at Queen's College, Galway, Ireland. Their assessment, published in 1866, was that *Eozoön* was "purely mineralic, nonbiologic." Not dissuaded, Dawson pressed on, turning to Karl Möbius (1825–1908), a professor of Zoology at Kiel, Germany, and arguably the world's expert on foraminifera, to whom Dawson sent innumerable specimens. After studying some 90 such specimens, Möbius' verdict was that they were "certainly not biologic." In Möbius' 1879 publication, this now being by then such a long-running 20-year-long controversy, he chided Dawson and those who had followed his lead: "No error will be changed into truth by constantly believing nor by persistently declaring it as truth."

Finally, in 1894, Dawson's long-claimed "biologic origin" of *Eozoön* was laid to rest (though he continued to champion this view until his death in 1899). Two English geologists, J.W. Gregory – who as a staff member of the British Museum [Natural History] had studied Dawson's specimens of *Eozoön* – and his colleague, Hugh Johnston-Lavis, discovered abundant specimens of this so-called ancient fossil in pressure-cooked limestones near Mount Vesuvius in southern Italy. As they first showed, we now know that Dawson's *Eozoön* is a non-biologically produced serpentinized metamorphosed limestone, the same type of handsome green-and-white marbled rock used commonly to floor US bank buildings, relegating Dawson to having the dubious distinction of being the first to describe and formally name a pseudofossil from the Precambrian rock record.

4.3 THE FIRST (ALMOST) BREAKTHROUGH DISCOVERIES: C.D. WALCOTT

The first to make lasting inroads into the solution of Darwin's dilemma, the founding hero of what would later become the science of Precambrian Paleobiology, was Charles Doolittle Walcott (1850–1927; Figure 4.4). Known most famously for his discovery in 1909 of the Cambrian-age Burgess Shale Fauna of British Columbia – a startlingly well-preserved diverse assemblage of early-evolved animals that form the basis of our knowledge of the "Cambrian Explosion of Life" – it is Walcott who, during the last two decades of the nineteenth century, made the initial prescient discoveries that in retrospect can now be seen to have set studies of Precambrian life on their paradigm-shifting course.

Like Logan and Dawson before him, Walcott was an influential "mover and shaker." Indeed, over his illustrious career, he served as director of the US Geological Survey; secretary (i.e., head) of the Smithsonian Institution; president not only of the US National Academy of Sciences but also of both the American Philosophical Society and the American Society for the Advancement of Science; and confidant and advisor of US president Theodore Roosevelt. In 1902, Walcott, along with several other prominent individuals, met with the "robber baron," steel tycoon and philanthropist Andrew Carnegie, to establish the Carnegie Institution of Washington as

FIGURE 4.4 Charles Doolittle Walcott.

a center for advanced research and training in the sciences. There is little doubt that had Walcott been a British citizen, he, like Logan and Dawson, would have knighted and known today as "Sir" Charles Doolittle Walcott.

Born in New York Mills in upper-state New York, and later a resident of Trenton Falls, from 1858 to 1868 Walcott attended the Utica public schools and Utica Academy from which he departed after the 10th grade, having never formally graduated. (In the mid-1800s, this lack of "complete schooling" was not uncommon, and for Walcott, later the recipient of numerous honorary degrees from colleges and universities in the United States and Europe, it turned out not to matter.) In his youth, Walcott developed an avid interest in natural history, collecting birds' eggs, minerals, and whatever fossils he came across. And after his family had moved to Trenton Falls in 1871, where Walcott had been hired as a farmhand, he began mostly collecting trilobites (a particularly rich collection, much of which two years later he sold to Professor Louis Agassiz of Harvard's Museum of Comparative Zoology).

In 1876, Walcott, a 26-year-old trilobite-collecting farmhand, began his professional scientific career when he was appointed as an assistant to James Hall (1811–1898), New York's chief geologist and state paleontologist. By all accounts, Hall was an irascible tyrant, but for Walcott, this science-based position was far better than continuing to till the fields. Moreover, it turned out to be pivotal for Walcott's later Precambrian discoveries.

A few years earlier, in about 1870, Hall had visited Saratoga Springs in the Adirondack region of New York State to examine concentrically layered originally mound-shaped structures in Cambrian-age limestones that had been first noted by John H. Steele in 1825 (Figure 4.5). Though the origin of these structures had been long debated, some arguing that they were merely inorganic concretions with

FIGURE 4.5 Cambrian-age *Cryptozoön* stromatolites, Saratoga Springs, New York.

others imagining them to be biological, Hall interpreted them to be algal reefs – structures now known as "stromatolites" – fossilized remnants of finely laminated microbial (predominantly cyanobacterial) communities that he formally named *Cryptozoön* (Gk., "hidden-life"). In 1878, two years after having signed on with Hall, Walcott visited the Saratoga site and also became convinced that the structures were biologic.

In 1879, after having spent only three years with Hall, Walcott joined the then newly formed United States Geological Survey (USGS) as a "geological assistant," employee number 20 at a starting salary of $50 per month. Later that year, he joined USGS director John Wesley Powell's first expedition down the Grand Canyon, a harrowing venture in the rapid currents carried out on unstable log rafts. In 1882 and again in 1883 he repeated this dangerous trek, in the latter year discovering algal reefs, like those at Saratoga, the first Precambrian stromatolites ever reported (from a site now known as "Stromatolite Ridge" of the 900-million-year-old Galeros Formation; Figures 4.6 and 4.7). Later, in 1899, Walcott found in other Grand Canyon rocks the first cellularly fossils ever recorded from the Precambrian, flattened disc-shaped millimeter-sized objects (*Chuaria circularis*; Figure 4.8), fossils he misinterpreted to be the compressed remains of tiny shelled animals that are now known to be remains of large single-celled planktonic algae.

At the beginning of the 1900s, Walcott's field work moved north along the spine of the Rocky Mountains, first to the Lewis Range of northern Montana – today, the Glacier National Park region – where he discovered numerous additional Precambrian stromatolites in which, after microscopic examination of the rocks following his return to Washington, D.C., he detected minute objects that he interpreted to be fossil bacteria. (Here Walcott had the right idea but this time the wrong

FIGURE 4.6 Precambrian-age Stromatolite Ridge, Grand Canyon, Arizona.

FIGURE 4.7 *Cryptozoön*-like stromatolite at Stromatolite Ridge, Grand Canyon, Arizona.

interpretation – some years ago, Bill had a good look at the slides containing Walcott's reported "fossil bacteria" archived at the US National Museum and found them to be non-biologic mineralic pseudofossils.)

From the US Rockies, Walcott then moved north to British Columbia, Canada, where he is credited as having discovered in 1909 the world-famous fauna of the Cambrian Burgess Shale. Over subsequent years, he formally described numerous

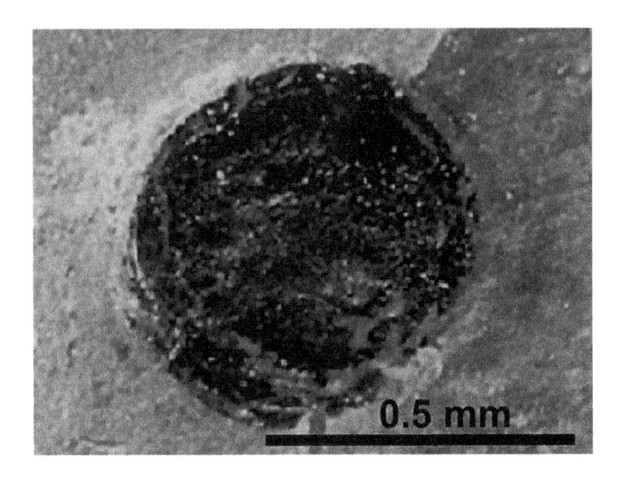

FIGURE 4.8 Walcott's *Chuaria circularis* (Precambrian Chuar Group, Grand Canyon, Arizona).

fossil animals previously unknown to science (e.g., *Wiwaxia*, a slug-like animal covered with scaly armor; *Ottoia*, a carnivorous worm up to 15 cm long; *Hallucigenia*, a velvet worm with needle-like spines; and many others [Figure 4.9]). He even named one of the new finds, the extinct arthropod *Sidneyia inexpectans* (i.e., "Sidney's surprise discovery") after his second-born son, Sidney Stevens Walcott, who later wrote of how he discovered this specimen on the first day of Walcott's 1910 exploration of the Burgess Shale site.

Walcott did it all! He was the first to discover Precambrian microbe-produced stromatolites (Grand Canyon, 1883); cellularly preserved Precambrian single-celled algae (*Chuaria circularis*, 1899) also from the Grand Canyon; numerous additional Precambrian stromatolites (Lewis Range, Montana, 1902–1907); and putative fossils

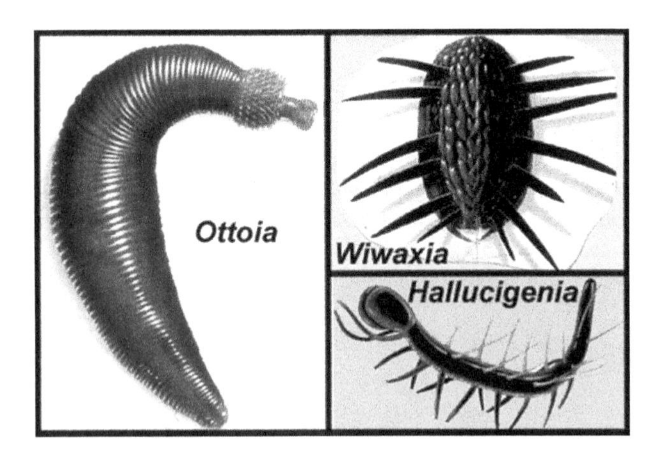

FIGURE 4.9 Models of three of Walcott's Cambrian Burgess Shale animals, British Columbia.

FIGURE 4.10 Charles Doolittle Walcott Medal of the US National Academy of Sciences.

of Precambrian bacteria (published in 1915). To honor his achievements, in 1934, the US National Academy of Sciences established the Charles Doolittle Walcott Precambrian Research Medal (Figure 4.10), the foremost accolade for contributions to such studies worldwide. And in 2009, to commemorate Walcott's 1909 discovery of the first evidence of the "Cambrian Explosion of Life," the Big Rock Brewery in Alberta, Canada, produced the limited-supply one-time-only Burgess Shale Centennial Celebration Ale. Remarkable! *Hooray for C.D. Walcott!*

4.4 WALCOTT MEETS HIS MATCH: A.C. SEWARD

Walcott's breakthrough discoveries had brought the solution to the so-called "missing" Precambrian life to the brink of success. Amazingly, however, this promising beginning soon turned to dust – partly because of unforced errors but most particularly because of the structure of early-1900s science and its reliance on "Science by Authoritative Assertion."

Walcott's Burgess Shale fossils were soon accepted – chiefly because his finds were consistent with what had been known before. But because of their non-dogma-supporting newness, his seminal finds of more ancient life were greeted with skepticism. Nevertheless, some workers were sufficiently impressed to enter the fray, carrying out field work, primarily in the Rocky Mountains of Montana, as a follow-up to Walcott's discoveries of Precambrian stromatolites. Unfortunately, however, the spate of reports stemming from this work (which extended into the 1930s) produced little more than hopeful but unsubstantiated erroneous claims – typically, of non-descript mud-chips and flakes exposed on bedding surfaces thought possibly-perhaps-maybe to be fossilized dispersed fragments of the carapaces of trilobites or similar ancient animals. Because none of these overly optimistic finds held up, these repeated failures cast a pall over the field.

At that time in the development of science, studies of fossil animals were carried out largely by Geology department–trained paleozoologists, well-schooled in the study of the fossil faunas of Cambrian and younger rocks but having no experience in the older Precambrian, little if any knowledge of fossil algae, and no acquaintance with fossil bacteria. In contrast, academic departments of Biology were then divided into two great subgroups, zoologists and botanists – there being no academic departments of Molecular Biology, or Genetics, or Microbiology and the like – and the study of fossil plants, the science of Paleobotany, was housed in the botany subset of Biology departments, not in departments of Geology.

Walcott's Precambrian discoveries centered on single-celled algae, supposed bacteria, and stromatolitic "algal reefs," all of which fell under the purview of Botany and, thus, of the world's leading paleobotanist, Sir Albert Charles Seward (1863–1941; Figure 4.11) who raised questions about – in fact derided, "mocked" – Walcott's findings.

Given Walcott's prominence and sterling scientific credentials, there is little doubt that his opinions could be cast in doubt only by another acknowledged "Great." Albert Charles Seward filled that bill. Professor of Botany at Cambridge University, England, Seward served also as the university's vice chancellor, the chief executive officer ("CEO") in charge of its day-to-day operations. Moreover, over his career, he served as vice-president of the Royal Society and president of the British Society for the Advancement of Science, and he was a fellow both of the Geological Society and the Linnean Society of London. And like Logan and Dawson decades earlier, Sir Albert Charles had been knighted by the British monarchy.

FIGURE 4.11 Sir Albert Charles Seward.

Seward was a prolific writer, in 1931 publishing his classic paleobotanical textbook *Plant Life through the Ages, a Geological and Botanical Retrospect* (Figure 4.12). In an age when the "sun never set over the British Empire," his volume was read worldwide. And in that volume, in his necessarily brief discussion of Precambrian plant fossils – at the time there being only scanty evidence to discuss – he refuted Walcott's acceptance of the biological interpretation both of stromatolites and fossil bacteria. Indeed, he authoritatively opined that Walcott's interpretation of "*Cryptozoön* is, I venture to think, not justified by the facts … It is clearly impossible to maintain that such bodies are attributable to algal activity." He then upped his criticism by asserting that "We can hardly expect to find in Pre-Cambrian rocks any actual proof of the existence of bacteria." And to this Seward added a bit of derisive, if creative, doggerel about Precambrian fossils: "Creatures borrowed and again conveyed, from

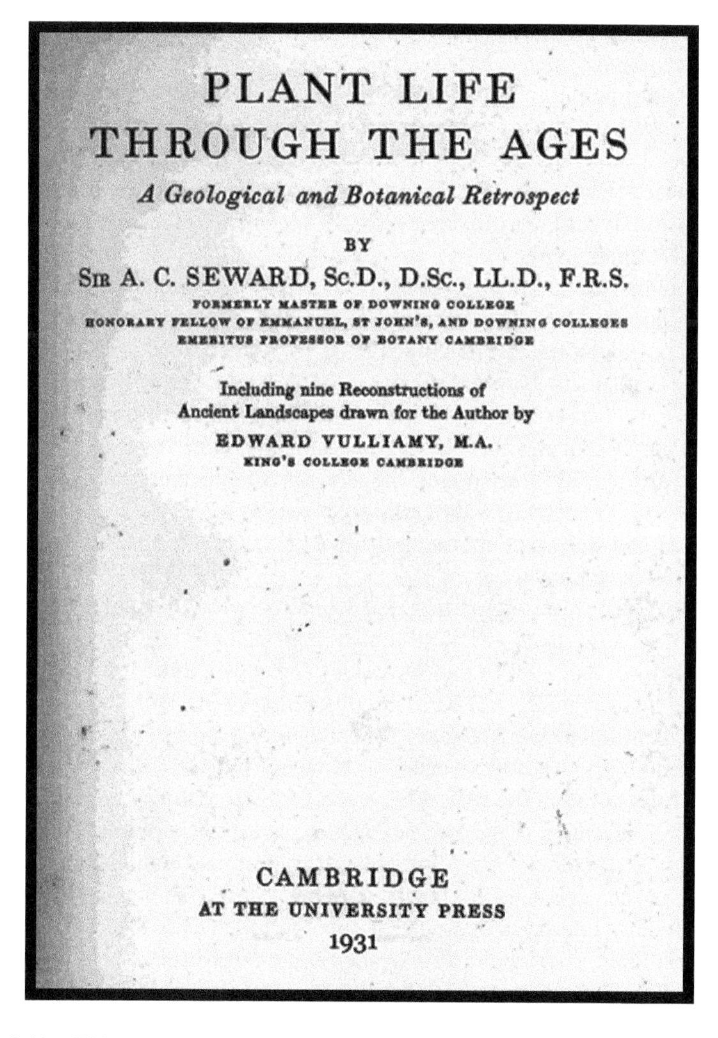
PLANT LIFE
THROUGH THE AGES
A Geological and Botanical Retrospect
BY
SIR A. C. SEWARD, Sc.D., D.Sc., LL.D., F.R.S.
FORMERLY MASTER OF DOWNING COLLEGE
HONORARY FELLOW OF EMMANUEL, ST JOHN'S, AND DOWNING COLLEGES
EMERITUS PROFESSOR OF BOTANY CAMBRIDGE

Including nine Reconstructions of
Ancient Landscapes drawn for the Author by
EDWARD VULLIAMY, M.A.
KING'S COLLEGE CAMBRIDGE

CAMBRIDGE
AT THE UNIVERSITY PRESS
1931

FIGURE 4.12 Title page of Seward's *Plant Life through the Ages.*

book to book – the shadows of a shade." (Walcott, having died three years earlier, could not respond.)

Seward, the acknowledged world-leading paleobotanist, had spoken. Progress ceased for the following three decades. Ultimately, his assessment was shown to be in error.

In the workings of science and the understanding of its history, facts are facts – like 'em or not! It is of course far easier to evaluate them in retrospect rather than when immersed in the culture of their time, but one can easily understand Seward's position: Walcott's evidence of life in the Precambrian was scanty; others had tried to back it but repeatedly failed; and Seward was simply following the long-accepted Darwin-derived dogma that evidence of early life was unknown and assumed unknowable.

It seems probable that the problem here was not so much the fault of Seward as it was of the contemporary scientific community. Seward spoke from a position on-high – authoritative "Science by Assertion" – and the community followed his lead. That can be detrimental to science. Yes, the authorities have knowledge and experience. But others have expertise too, and their opinions also matter, especially if the acknowledged leaders unthinkingly follow accepted dogma. In short, it seems apparent that knowledge will best advance if each and every worker personally evaluates and carefully analyzes the available evidence. Uncritical reliance on the opinions of the "Greats" is anathema.

To Bill Schopf, this matters because he is concerned about the present-day students. It is important for them to think for themselves, to individually analyze the available facts – and if the facts seem not to fit, to wonder, doubt, raise questions, and "suss out" (a Britishism) a more plausible solution. Moreover, with regard to Seward, Bill's friends at London's Natural History Museum (especially those familiar with its Seward Paleobotanical Library) have occasionally chided him for his criticism of the role Seward played in stymieing the advance of Precambrian paleobiology for a span of nearly 30 years. Seward certainly deserves his due – he was exceedingly accomplished and was no doubt doing the best he could – but Bill is convinced that Seward should have known better.

Why would Bill think such a thing? It would be wrong to fault Seward over his doubt of *Cryptozoön*, stromatolitic structures that had been debated in the geological literature since the early 1800s and were not to be shown until 1961, and the discovery of modern stromatolites, to be assuredly biologic. But Bill does take exception to Seward's dismissive assertion that "We can hardly expect to find in Pre-Cambrian rocks any actual proof of the existence of bacteria." Indeed, Seward's writing shows that he was fully aware of the 1917–1921 series of five lengthy monographs by Robert Kidston and William H. Lang describing the plant fossils of the Early Devonian Rhynie Chert of Aberdeenshire, Scotland – the oldest fossil land plants then known. And from that he would have known that in the last volume of this series, they reported cellularly preserved exceedingly tiny microbial cyanobacteria, since shown to be the dominant microorganisms of the photic-zone Precambrian biota and the principal builders of stromatolites (their preservation in this 400-million-year-old deposit being not surprising, given that cyanobacteria, like higher plants, have robust cell walls).

4.5 BEGINNINGS OF THE RENAISSANCE: 1950s

Finally, in the mid-1900s, slowly but surely the tide began to turn; the field began to awaken. In 1947, the Australian geologist Reginald C. Sprigg (1919–1994) reported his discovery of imprints of fossil jellyfish in the Pound Quartzite of the Ediacaran Hills of South Australia, strata that he regarded as being either lowermost Cambrian or, possibly, uppermost Precambrian. A few years later, these finds were shown by paleontologist Martin F. Glaessner (1906–1989; Figure 4.13) and his co-worker Mary J. Wade (1932–2005; Figure 4.14) to date from the very latest Precambrian, for the first time establishing the existence of Precambrian animals. Also in the 1950s, Boris V. Timofeev (1916–1982) and his colleague Tamara N. Germann (Hermann) in Leningrad, USSR, published the first of what were to become numerous detailed studies of microscopic planktonic plant fossils compressed in Precambrian shales of the Soviet Union, most notably from Siberia, important discoveries that at the time were little known in the West. And in the United States, in 1942 Preston E. Cloud (1912–1991; Figure 4.15) authored one paper on Precambrian stromatolites; and, in 1954, Stanley A. Tyler (1906–1963; Figure 4.16) and Elso S. Barghoorn (1915–1984; Figure 4.17) published a short note reporting the discovery of microbial fossils in the mid-Precambrian Gunflint chert of Ontario, Canada.

Though the field had begun to stir, it was not until the decade of the 1960s that the seminal discoveries received wide notice, most prominently from Glaessner's article in the March 1961 issue of *Scientific American* which featured an Ediacaran Precambrian animal fossil on its cover, and from a sequence of three articles in the journal *Science* in 1965: first in February (a Barghoorn-Tyler paper describing the Gunflint microflora); then in April (Cloud's contribution,

FIGURE 4.13 Martin Fritz Glaessner.

FIGURE 4.14 Mary Julia Wade.

FIGURE 4.15 Preston Ercelle Cloud, Jr.

FIGURE 4.16 Stanley A. Tyler.

FIGURE 4.17 Elso Sterrenberg Barghoorn.

FIGURE 4.18 J. William (Bill) Schopf.

validating the Barghoorn-Tyler report); and lastly in October, a Barghoorn and Schopf (Figure 4.18) paper reporting Precambrian microbes from the Bitter Springs chert of central Australia, fossils decidedly better preserved than those of the older Gunflint deposit, which sealed the case by showing once and for all that the Gunflint fossils were no fluke, that a cellularly preserved Precambrian record of life actually exists.

4.6 DISCOVERY OF THE MID-PRECAMBRIAN GUNFLINT FOSSILS

Discovery of the 1,900-million-year-old microbial assemblage of the Gunflint chert of Ontario, Canada – a game-changing find at the beginning of the paradigm shift from the unknown to the known early record of life – dates from 1952 and the prescient curiosity of Stanley A. Tyler of the University of Wisconsin. Tyler, an accomplished economic geologist, was mapping and studying the mineralogy of the Gunflint Iron Formation that rims the northern edge of Lake Superior, at that time a major source of taconite iron-ore used to produce steel. His studies took him from Duluth, Minnesota, at the far west of the Gunflint outcrops, to the eastern margins of Lake Superior in southern Ontario. One Sunday, toward the end of that field season, he rented a boat and an outboard motor at the fishing village of Rossport, northeast of Thunder Bay on the lake's northernmost shore, and motored south toward Schrieber to get in a day of fishing. Around noon, he found himself opposite Flint Island and, as luck would have it, noticed an unusual black outcropping of the Gunflint rocks on the nearby Canadian coast where he beached his boat to have a look.

The rocks were indeed rather odd – relatively unmetamorphosed (only slightly "pressure-cooked") fine-grained black carbonaceous cherts (chemically precipitated siliceous quartz-rich rocks) occurring as concentric rings in meter-broad flat-topped and planed-off circular structures (originally mound-shaped stromatolites, the tops of which had been ground away by glaciers during the Pleistocene Ice Age; Figures 4.19 and 4.20). During his earlier work to the west, Tyler had seen occasional chunks of

FIGURE 4.19 Gunflint chert outcrop at the Schreiber Beach Locality.

FIGURE 4.20 Gunflint stromatolites at the Schreiber Beach Locality.

such black cherts, most commonly occurring as broken bits and pieces embedded in the bright- to deep-red iron-rich rocks characteristic of the unit. However, this exposure of the Gunflint, the "Schrieber Beach locality," was unlike the other outcrops he had studied. He gathered a few samples and returned to his fishing.

When he returned home to his university at Madison, Wisconsin, he had specimens from his collections sliced, ground, and mounted on glass slides (petrographic thin sections) so that he could study their mineralogy by optical microscopy. Because the Schreiber black cherts lacked the iron minerals that he was investigating, they initially held not much interest. But by the end of that summer – for completeness and out of curiosity – he studied them as well and discovered much to his surprise that they were packed with microscopic dark-brown fine thread-like strands, sporadically peppered by similarly brown-colored ball-shaped to ellipsoidal and seemingly hollow spheroids (Figure 4.21).

Though Tyler was an economic geologist, not a paleontologist, he thought to himself, "My Goodness! These certainly do not look like minerals – perhaps they are microscopic fossils." He took down from his office shelf his one paleontology textbook, the classic 1935 Shrock and Twenhofel *Principles of Invertebrate Paleontology*, to see if he could find anything like them. He didn't – the volume focused on invertebrate animals, not minute fossil microbes (which at the time were essentially unknown). Still, all was not lost. That year, 1952, the annual fall meeting of the Geological Society of America (GSA) was to be held in Boston, Massachusetts, and Robert R. Shrock (1904–1993), lead author of the text, was a professor at the Massachusetts Institute of Technology (MIT) in nearby Cambridge.

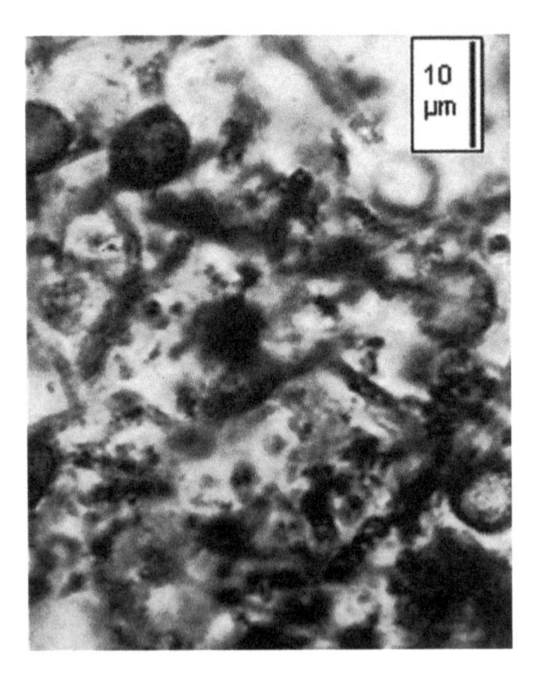

FIGURE 4.21 Gunflint chert microfossils.

Perhaps Tyler could locate Shrock at the meeting, show him some photographs, and get his opinion.

In early autumn, Tyler journeyed off to the 1952 GSA meeting. The conference was huge, attended by some 6,000 geologists from around the world. Though it would be difficult to locate Shrock, Tyler had a plan – his best chance to find Bob Shrock was at the annual MIT Tuesday evening cocktail party, that year to be held at the Parker House Hotel – which he then attended and found Shrock along with his colleagues and former students. After outlining the situation, Tyler showed Shrock photos of the microscopic objects. As Shrock years later told Bill Schopf, his response was immediate and positive: "They look just like the fungi at the top of a jam-jar that has been left open too long." Wow! Tyler's notion was now confirmed by a world-class expert. Tyler and Shrock were both excited – as best they knew these were the first "real" Precambrian fossils ever discovered!

But Shrock continued, "Understand, however, I am not an expert on fungi. But I know who is – your friend and mine, Elso Barghoorn [the paleobotanist at Harvard]. Let's find him and see what he thinks." (Shrock was aware that Tyler and Barghoorn had previously teamed up on a previous study of the Michigamme Coal of northern peninsula Michigan and, more importantly, he also knew that during World War II, Barghoorn had been stationed in Panama, where he investigated the tropical microscopic fungi that infested the leather carrying-cases of binoculars used by American forces in the Pacific theatre. Here, yet again, serendipity – happenstance "luck" – played a role in furthering the science.)

Together, Tyler and Shrock departed the MIT cocktail party and headed off to the Harvard party to find Barghoorn, only to encounter him in the Parker House lobby as he was on his way to the MIT gathering. There, in the hotel lobby, the three sat, chatted, and discussed Tyler's find. Barghoorn, too, was convinced. This was a breakthrough discovery!

Upon his return to Madison, Tyler gave his graduate students the task of photographing the best fossils they could find. He and Barghoorn then submitted a short note to *Science*, published in April 1954. To discourage others from mining the outcrop before they had finished their work, they purposefully neglected to specify the exact site from which the fossils were obtained, noting only that it was "near Schreiber, Ontario." And though they misidentified the most abundant of the fossils as filamentous "fungi" (a result of Shrock's suggestion and Barghoorn's predilections) and erroneously described others as "probable calcareous flagellate protozoans similar to modern *Discoaster*," they correctly recorded the occurrence of two types of microscopic "blue-green algae" (cyanobacteria). This brief eight-paragraph announcement paper represented the first decisive stride along the path that set the paradigm-changing field of Precambrian Paleobiology on its present course.

4.7 BILL'S INITIAL INVOLVEMENT IN THE GUNFLINT STUDIES

Bill Schopf's entry into doing "real science" in the field dates from the early summer of 1961, at the end of his second year of college. After he had written to both Cloud and Barghoorn expressing his interest in uncovering the "missing" Precambrian record of life – to which both kindly responded – Bill realized that crux of the problem was to discover actual fossilized organisms (cf. the Tyler-Barghoorn 1954 announcement paper), not simply the megascopic stromatolitic structures they were alleged to have produced (cf. Cloud's 1942 paper). He therefore wrote a letter to Professor Barghoorn asking his permission to come to meet him at the end of that school year. Bill has no idea what Barghoorn thought about this bold request – it is quite plausible that the Harvard professor figured that Bill was some sort of brash, wild-eyed, know-nothing second-year college student who was simply going to waste the learned professor's time. Still, because in his pre-Harvard days Barghoorn had taught at Amherst College, he was likely to have known something about Oberlin College and was likely to have heard of Bill's father, James M. Schopf, a fellow paleobotanist. In any case, Barghoorn agreed to Bill's request.

In retrospect, one can suspect that Barghoorn actually had little choice but to grant this audience. Bill's impression is that no one before him may have evinced a deep interest in tackling this problem – but Bill had no way to know. What he does now know is what Preston Could later (in 1983) wrote about his participation in what Cloud termed the "vanguard" of the field: "J. W. Schopf was the first person to begin at early age to prepare for a career in this field, and in that sense the first to be adequately prepared for it." If that is really true, it strikes Bill as amazing. Ask yourself: "How it could be that the Tyler-Barghoorn announcement paper seven years earlier did not elicit more interest?" That seems odd, though evidently Cloud was right (or, more correctly, Bill knows of no counter-examples). *Once again, Bill seems to have been plenty lucky!*

During Bill's 1961 visit, Professor Barghoorn was "sizing him up." OK by Bill – he had been permitted to meet with an authentic "Great." But what Barghoorn might not have realized was at that same time Bill was "sizing him up" and wondering to himself whether he could meet the demands of four or five years with this professor and, also importantly, whether he could stomach becoming a member of what Bill imagined to be the Harvard "better-than-thou" milieu (about which he had serious misgivings). In each other's eyes, they both passed these tests. Toward the end of that visit, Barghoorn handed Bill a small chunk of the microfossil-bearing Gunflint chert – the subject of Bill's subsequent Oberlin Honors work – thinking, one may now suppose, "What will this well-meaning upstart manage to do with this rock?"

As it turned out, during Bill's Honors work, he found innumerable good fossils, including both those that had survived more-or-less intact in the microbial mats they produced as well as much less common specimens of planktonic microbes that had filtered down into the deposit from overlying waters, some of which were unlike living microorganisms and particularly bizarre. For example, one was composed of a central medium-sized ball covered by scattered smaller ball-shaped cells, all enclosed in a larger organic-walled spheroid (Figure 4.22) – rather like a volleyball covered by golf balls enclosed by a basketball (Figure 4.23) – a fossil that Barghoorn later named *Eosphaera tyleri* ("Tyler's dawn spheroid," its species name honoring Stanley Tyler, the discoverer of the Gunflint fossils). And another bizarre fossil featured a basal ellipsoidal cell connected by a slender tube to a larger overlying umbrella-like crown (Figure 4.24) – a form that reminded Bill of a chocolate-covered peanut subtending a cocktail party–sized Japanese umbrella (Figure 4.25) – later named *Kakabekia umbellata* (for southern Ontario's Kakabeka Falls, where

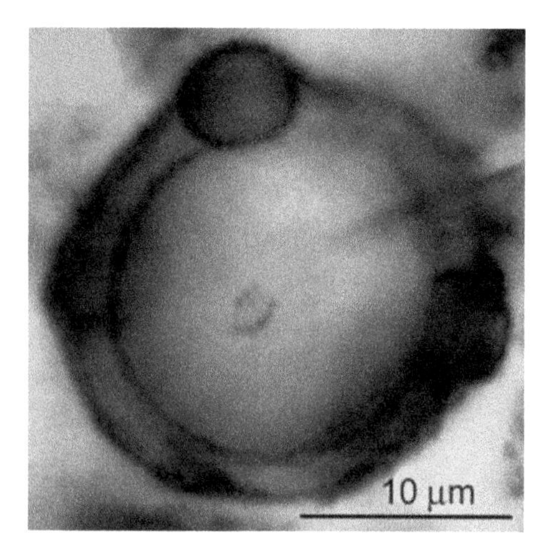

FIGURE 4.22 *Eosphaera tyleri,* Gunflint chert.

FIGURE 4.23 A fanciful model of *Eosphaera tyleri.*

this form occurs in abundance, and for its prominent umbrella). At the time, these and other microscopic fossils included in Bill's Honors thesis were new to science, though most (but not all) have since been found in other deposits of about the same age (2,100–1,800 million years) and are now regarded as forms of life that ultimately became extinct (like trilobites and dinosaurs).

4.8 BILL'S UNDERSTANDING OF THE BEGINNINGS OF THE FIELD

Much of the foregoing pre-1960 history of this science Bill learned during his undergraduate days at Oberlin College from his repeated library forays when he sought to understand whether his quest to uncover the "missing" record of life was

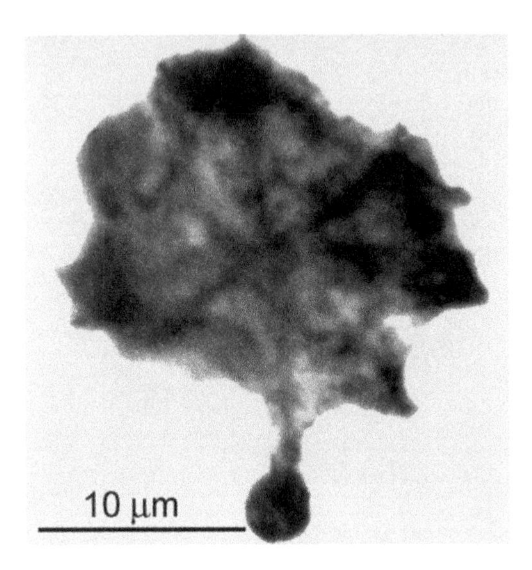

FIGURE 4.24 *Kakabekia umbellata,* Gunflint chert.

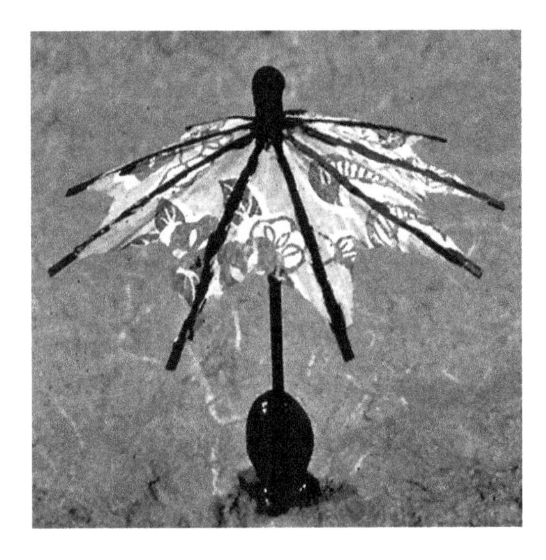

FIGURE 4.25 A fanciful model of *Kakabekia umbellata.*

a worthwhile venture – with aspects of the "oral history" of the Gunflint-discovery tale having been outlined to Bill by Barghoorn, first during Bill's 1961 visit and the gaps filled in later by Barghoorn and MIT Professor Bob Shrock during Bill's graduate-school days.

During his college years, Bill became increasingly committed to his quest, due largely to great fun he was having in his (non-faculty-guided) Honors research project. And, as he had learned from his forays to the Oberlin library, Darwin stated the problem, Walcott showed that the answer could be found, and the heroes of the soon-to-emerge new science were Sprigg, Glaessner, and Wade, in Australia; Timofeev and Germann in the Soviet Union; and Tyler, Barghoorn, and Cloud in the United States. Bill later came to know all but three (Sprigg, Wade, and Tyler) of these field-founding "movers and shakers."

But this narrative is now getting ahead of itself. Let us now return to Bill's graduate-school days (1963–1968) to show how Precambrian paleobiology then began an almost immediate up-swing to its current status as an established international and interdisciplinary paradigm-changing field of science.

5 Graduate School and Early Career

5.1 FIRST YEAR OF GRADUATE SCHOOL

In June 1963, immediately after graduation from Oberlin as a fortunate and appreciative "High Honors" Geology student, Bill headed off to Harvard to study in Professor Barghoorn's lab, where, as a Biology student, he imagined that he would expand on his undergraduate work and do a doctoral thesis on the Gunflint microbiota. Six other grads from Bill's Oberlin class were to join him that fall, and one of his tasks that summer was to rent a house near the campus for the seven of them. He found an appropriate house four blocks from campus and made arrangements to rent it beginning 1 September. Bill totted up his resources and decided that he could handle the required pre-payment of the first month's rent, especially since he had an NSF Fellowship due to begin that month (such graduate student fellowships being in those days no doubt far easier then to obtain than they are at present).

Ahh ... almost but not quite! As it turned out, though the NSF Fellowship was to begin September 1, it would not provide Bill with funds until September 30. He was $250 short on the required rent! What to do? Bill was not about to ask his parents for a loan – he was independent, on his own, and it was a responsibility that he alone must meet. So, Bill went to the local Cambridge bank and asked for a one-month $250 loan. As it turned out, their lowest loan was $500 to be repaid within six months. So be it. Bill agreed, signed the papers, got the loan, and repaid the loan six weeks later, in early October after the NSF money arrived, glad to have taken care of this matter without parental help.

Thereafter Bill never thought much about this. Then, a year later he again found himself in a financial bind. Professor Barghoorn, his soon-to-be new wife (née Dorothy Osgood), and Bill embarked on a field trip heading from northern peninsular Michigan in the south (Nonesuch Shale, Michigamme Coal), west to Duluth Minnesota (Sudan Mine, Biwabik Iron Formation), then east along the northern shore to Lake Superior to the Schreiber Beach locality (Gunflint chert). They finished their work four days early, ending up in Val-d'Or, Quebec, when it occurred to Bill that his older brother Tom would be receiving his Ph.D. from Ohio State only a couple of days later. Their parents were away in Edinburgh, Scotland, attending the 1964 International Botanical Congress and would not be present for Tom's big day. And Tom's wife Carrie, a physician, was stuck "on call" at Massachusetts General Hospital in Boston. But, if Bill could get a plane from Montreal to Columbus, at least he could be there to cheer on older brother Tom.

Bill's problem was that he didn't have enough money with him to pay the plane fare. He went to a Val-d'Or bank to inquire about a short-term loan, and they then called his Cambridge bank to check his *bona fides*. Remarkably, the loan officer with whom Bill had dealt a year earlier was now the bank branch manager and, even more remarkably, the bank manager remembered him! Asking Bill how much money he needed and adding a couple hundred dollars to give Bill extra lee-way, the manager immediately authorized a "no-cost loan" and arranged with the Val-d'Or bank to provide the funds. Bill got to Columbus, saw Tom graduate, returned to Cambridge where he had some cash put aside, and repaid the total loan the following Monday. *Gee Whiz Gosh – Bill was lucky once again!*

During Bill's first year in graduate school and the following summer, he spent most of his research time working on the 1,000-million-year-old Nonesuch Shale of northern peninsular Michigan, during which he learned a fair amount of organic geochemistry and the rudiments of carbon isotopic geochemistry – both completely new to him – a project led by Barghoorn and organic geochemist Warren G. Meinschein (then a research associate at the Esso Research and Engineering Company, Linden, NJ). The product of this venture was the first truly interdisciplinary contribution to Precambrian Paleobiology, combining aspects of Geology, Micropaleontology, Biomarker Organic Geochemistry (porphyrins and hydrocarbon side-branch derivatives of chlorophyll), and Carbon Isotope Geochemistry (measurements of the isotopic composition of fossilized carbonaceous organic matter). The study was regarded to be sufficiently innovative to merit two papers in *Science*, one in 1964 and the second as a lead article in 1965 – not bad for Bill's first two co-authored publications as a fledgling graduate student!

During that year, Bill passed his foreign-language exams (French and German), a proficiency then required of doctoral science students, and at the end of the year he weathered the Biology department's "up or out" first year of progress decision (those not measuring up being awarded a terminal Master's degree).

Bill also learned a good deal about his professor's history. Barghoorn had been a star graduate student of Harvard's internationally known botanist Irving Widmer Bailey, after which he took a teaching job at Amherst College. When the previous Harvard paleobotanist departed (under a cloud due to alleged plagiarism) Barghoorn was brought back as a member of the Harvard faculty. For the 10 years prior to Bill's entry as his graduate student, Barghoorn had experienced a very tough time: two divorces, the suicide of his eldest son, and a bar-fight in the red-light district of Amsterdam that damaged one eye; back in Cambridge, he had suffered through a broken leg and painful kidney stones. He coped with these difficulties by drinking alcohol to excess. His rumored behavior with Radcliff College co-eds was a subject of some gossip (in those days, such behavior being not particularly uncommon for senior faculty); he was continuously obsessed about being "scooped" in science by his perceived rivals; and at one point, he told Bill that he had not received a pay raise for the 10 years before Bill came work with him (showing Bill his required IRS tax report – ostensibly to complain about his tax burden – and telling Bill that he had "finally begun to get [his] due").

Politically, Barghoorn was an arch conservative (in 1964, purposefully leaving town on election day so that he would not have to own up to his colleagues that he had voted for Republican Barry Goldwater, of which they would not have approved. And in scientific articles, Barghoorn refused to reference any publications, however relevant (like those of Timofeev), authored by Soviet scientists, who were in his term

"Godless Commies." It was clear to Bill that his professor had come to view himself as a second-tier member of the Harvard Biology faculty (primarily, Bill guessed, because 11 were members of the National Academy of Sciences and he was not). Except when Barghoorn was teaching his one course per year, he spent most of his time at his farm in Carlisle, northwest of Boston, coming to the lab for two- or three-hour stints typically twice a week.

To an outsider, Professor Barghoorn would have seemed nowhere close to being an "ideal professor." But for Bill he was. He was enormously intelligent and knowledgeable; he instinctively identified excellent scientific questions, however difficult to solve; he had the largest vocabulary of anyone Bill had ever known; and he trusted Bill enough to let him roam freely, to "do his thing." Bill respected his professor immeasurably. Barghoorn was kind, considerate, and generous to Bill – he was a scholar and mentor from whom Bill learned an enormous amount, and Professor Barghoorn treated Bill like a son.

When Bill returned home for spring recess during his first year of graduate school, Bill's father ribbed him about being at Harvard: "You can always tell a Harvard man ... but you can't tell him much!" And, being a paleobotanist, as was Barghoorn, Bill's father asked him "How is it to work with Barghoorn?" Without mentioning Barghoorn's checkered past, Bill replied with supportive enthusiasm, listing off Barghoorn's many sterling traits.

As Bill has suggested to his students on countless occasions, "No one is perfect, me included!" That was a great lesson he had learned from being a student of Professor Barghoorn. And to Bill, the upshot of that lesson was that he and his students should pick and choose from the various traits of persons they admire – emulate those that seem valuable but ignore those that are not helpful.

5.2 GOOD STUFF AND A CRISIS

To be closer to the lab, during his second year in graduate school Bill lived in the attic of a Harvard-owned dilapidated abode half-a-block away, shared a bathroom with eight others, and cooked his evening meal on a hot-plate at the lab after his lab-mates had departed. The price for his lodging was right – $30/month. *Good stuff!* A year later, in 1965, Bill married his Oberlin girlfriend, Julie Morgan, daughter of a Mt. Holyoke Political Science professor. *More good stuff!*

During that period, the Viet Nam War was surging onward, but for the time being, Bill had a student deferment. Later, however, more soldiers were needed, a compulsory draft was instituted, and Bill, along with many thousands of others, took a pre-induction physical exam. His plan, if drafted, was to be assigned to Fort Detrick, Maryland, where he could learn hands-on microbiology. Bill then received a letter from his Columbus, Ohio, draft board informing him that for the upcoming draft-period, his birth date was ranked at 273 (out of 365), and that the previous year's draft had reached down only to 180; and urged Bill, "in [his] best interests," to enter his name in the pool of potential draftees. Bill did. He was not drafted. But he had met his responsibility. (When Bill later returned home to Columbus, he visited the draft board and personally thanked them. Heck, they had been enormously kind to him! They then explained to Bill that they "just didn't have too many potential draftees of his [supposed] caliber" and they would rather have him stay in school than to be, in their terms, "cannon fodder." For him, yet more good stuff! *Bill was lucky, yet again!*)

Barghoorn's plan for the summer of 1964 had been to finally (after his lost decade) prepare a definitive paper describing the 1,900-million-year-old Gunflint microbiota (the subject of the Tyler-Barghoorn announcement note of 1954 and of Bill's Oberlin Honors thesis). That summer, however, was consumed by work on the Nonesuch Shale study, so the Gunflint project (which Bill had previously assumed would be the focus of his doctoral thesis) was shelved for the time being – just as it had been by Barghoorn for the previous decade. Eight years earlier, in 1956, Tyler had provided a nearly completed lengthy manuscript to which Barghoorn had only to add taxonomic descriptions – a manuscript that since its arrival had languished on Barghoorn's desk. Both he and Tyler had assumed that once Barghoorn got around to doing his part of the job, Tyler would be the lead author of this landmark work. Unexpectedly, however, in the late summer of 1964, Tyler died (at the age of 57). Barghoorn journeyed to Madison, Wisconsin; obtained all of Tyler's notes, films, and fossil-containing petrographic thin sections; and turned them over to Bill to sort out.

In mid-autumn of 1964, when Bill was about half-way through the arduous task of relocating the tiny specimens photographed by Tyler's students (for which Bill had no information other than the slide-number of their thin section in which they occurred, intermeshed among tens of thousands of other minute specimens), a crisis emerged. Preston Cloud, Barghoorn's nemesis, had located the site of the original find, collected samples, studied the fossils, and had submitted a lengthy manuscript to *Science*, which had been sent to Barghoorn for review. Upon its receipt, Barghoorn charged into Bill's office, where he was assiduously trying to relocate the tiny Gunflint specimens, and loudly exclaimed: "What are we going to do?" (Barghoorn's use of "we" in that exclamation having understandably stuck in Bill's memory for all these years).

Barghoorn's' worst fears had materialized – because of his lethargy and personal difficulties, he had been "scooped," even though the 1954 Tyler-Barghoorn note had craftily not revealed the exact location of the find. Yet Cloud, a first-rate geologist-invertebrate paleontologist had ferreted it out (Cloud telling Bill later that a rainstorm had trapped him on the Lake Superior shoreline outcrop, accessible only by boat, and that he spent that extra time mapping the outcrop – a presumably quite useful map that, as best is known, Cloud unfortunately never published).

Given the emergency posed by Cloud's manuscript, Barghoorn instructed Bill to cut his classes for the following two weeks and, as Barghoorn worked on his contribution to Tyler's yet-unfinished manuscript, Bill was to prepare the photographic images and help him on the taxonomic descriptions of the Gunflint microbiota. Bill did as he was told.

The photographic work was a formidable task. Tyler's cache contained innumerable rolls of film taken by his students over several years, only a few of which included their original labels. And though Bill had done photographic work for his Oberlin Honors thesis, the facilities at Harvard were completely different. His tasks were to get in the dark room, mix the chemicals, prepare the trays, scan the films, select images of the "best" specimens, and print, dry and then present the pictures to

Barghoorn – nitty-gritty aspects of which Bill had not done since the 7th grade! But he was not about to admit his inexperience to his professor. Bill got out a couple of books, refreshed his memory, and got to work.

For the following two weeks, each day Bill located the best images and each night then printed pictures so that by the next morning he had organized the specimens into taxon-like categories that, based on his Oberlin Honors work, seemed sensible to him. Barghoorn would then arrive, review Bill's work, and prepare taxonomic descriptions of the various categories Bill had selected, by now all more properly identified as fossil cyanobacteria or as being of uncertain affinities rather than "filamentous fungi" or "probable flagellate protozoans" as the Tyler-Barghoorn 1954 announcement paper had suggested (changes perhaps at least marginally influenced by their descriptions in Bill's Honors thesis, which he had sent Barghoorn upon its completion two years earlier). Toward the end of each day, Barghoorn would give Bill a list of the new Latin names that he had selected (Bill's high-school Latin being better than his), which Bill would then check with Dr. Leslie Garay, Bill's orchidologist friend down the hall and a first-rate taxonomist (whom Barghoorn disliked, regarding him to be a "closet homosexual"). Each following day, Barghoorn would check the (now Garay-revised) names and move on to the next categories. So it went, day after day.

Finally, after some two weeks, the job was (mostly) done. Barghoorn then called Pres Cloud. Bill stood beside his office desk as he made the call. "Say, Pres, I received your manuscript to review. But my own manuscript is now ready to submit." Cloud's rejoinder: "Seeing is believing!" – which Barghoorn then repeated to Bill, his hand cupped over the phone. This back-and-forth banter went on for quite a while – with two or three more repeated "Seeing is believing" comments – until Cloud agreed that if Barghoorn actually had such a manuscript, he, Pres Cloud, would permit Barghoorn to publish first. The next morning, Barghoorn's manuscript and prints of Bill's still unfinished (and at that point, embarrassingly poor) plates were sent to Cloud.

Once given the go-ahead by Cloud, Barghoorn then confronted the problem of the order of authorship. He asked how Bill thought he should handle this. Bill's response: "You should be first author. Professor Tyler is no longer alive and someone needs to take the responsibility." Barghoorn seemed pleased, this being apparently what he wanted to hear. He then asked Bill: "Would you like to be a co-author?" Bill responded, "No. I have not done enough." That also seemed to please him.

Over the years, various people have questioned Bill's decision not to accept co-authorship of this work: "Without you, it would not have happened" – "It could have made you famous" – "You passed up a golden opportunity." Even Bill's father was skeptical, telling him later that Bill's work "got Barghoorn elected to the National Academy." Well, Barghoorn did soon get elected – at which time Bill went out and bought a bottle of champagne for the lab. (But, as Bill had not foreseen, their lab "celebration" was not as wondrous as he had envisioned, Barghoorn's election being primarily important because it relieved him of his perceived stigma of being regarded as a second-tier member of the Biology department faculty.)

Still, to Bill's mind, the well-meaning skeptics of his decision not to co-author the Barghoorn-Tyler breakthrough contribution simply did not understand the situation.

The fact was that by that time, Bill knew full well that this Gunflint work really mattered to his professor – after all, this was Barghoorn's very best chance to make history, to have his "time in the Sun." And Bill was but a mere second-year graduate student. It was not his place to detract from Barghoorn's forthcoming fame. That was fine with Bill – if he were ever to deserve such recognition, his time would come.

In subsequent years, others have also criticized Bill – though never to his face – for his cooperation (i.e., his "complicity") in helping Barghoorn prepare this manuscript, viewing Bill's participation as inappropriate or, to some, even "ruthless." However, Bill has never so regarded it and chalks up such reactions to ignorance of the priority of the find established by the Tyler-Barghoorn 1954 paper announcing the discovery and of professor–student relations throughout academia, but especially at Harvard, during the 1960s. In those days, Harvard professors were above reproach, the university being widely regarded as the foremost in the world and its faculty as among the very best. And as a mere graduate student, Bill did as he was told and could not plausibly have done otherwise. Pres Cloud, Bill thinks, deserves the credit for finally spurring Barghoorn into action and, also, for recognizing that priority for the discovery properly belonged to Barghoorn and the discoverer of the deposit, Stanley Tyler.

In some respects, this episode is reminiscent of the Darwin-Wallace discovery of evolution by natural selection, Darwin having a 15-year priority for the insight, helped ultimately by Lyell and Hooker, with Wallace receiving only "Honorable Mention." In this case, Tyler and Barghoorn had a 13-year priority for discovery of the Gunflint fossils – though Tyler died before he could bask in the glory of his find – and Bill, for Barghoorn, played the Lyell-Hooker "helper-role" with Cloud's Wallace-type contribution being that of the "Honorable Mention Validator." Those are the facts. Science sometimes is neither crispy clean nor interpersonally pretty – not the ideal enterprise that one might prefer – but the facts deserve to be recorded.

After Barghoorn had resolved the questions regarding authorship of the Barghoorn-Tyler paper soon to be submitted, he gave Bill the Cloud manuscript to review, which Bill did as thoroughly as he could. As it turned out, Cloud had made numerous errors – not his fault; he was an invertebrate paleontologist and had limited knowledge of the microbial world. When Cloud's paper was published, subsequent to the Barghoorn-Tyler landmark paper, Bill discovered that everything he had suggested to improve the presentation had been summarily ignored. After that – quite oddly, Bill thought, because Cloud would have had no way to know that it was he, not Barghoorn, who had reviewed his manuscript – Bill was told by their mutual acquaintances that Cloud regarded Bill to be his "biggest competitor." To this day, Bill does not know why. He was only a second-year graduate student! (And, anyway, it seems to Bill that in science, seizing on the notion of having "competitors" – in Barghoornian terms, "rivals" – is anathema. It's the evidence, the firm facts, not the presenter of those findings, that matter.)

During this time, Bill was the sole Biology student enrolled in a Geology department seminar course addressing the then relatively new notion of "continental drift" (plate tectonics). The course was taught by sedimentologist Ray Siever ("pro-drift") and paleontologist Bernie Kummel ("anti-drift") and concluded with a book review–like term paper in which the students were to evaluate a small volume that included eight or ten papers arguing differing sides of the question. After the course grades were given, the other six students in the class descended on Bill's office and demanded to know why he had received an "A" and they all received a "C" – though Bill had no idea how they knew his grade; he didn't know theirs. Evidently, the other students had reviewed each paper in the volume one by one, whereas Bill had grouped the papers into categories and reviewed the book as a whole (which, in fact, was the class assignment). A couple of the students were particularly irate (a grade of "C" in a Harvard graduate course being close to lethal), with one asserting to Bill: "Barghoorn helped you on this, didn't he?!" That was not so. Barghoorn knew absolutely nothing about either the course or Bill's essay.

At the time, Bill was dismayed and rather shocked. This was a far cry from the mutual support and appreciation he'd grown used to at Oberlin. But in retrospect, he now thinks he may now understand. To wit, he imagines that the students were particularly miffed that a mere Biology student, a member of the "unwashed Life Science tribe," had evidently outperformed members of their "pure, first principle-based Physical Sciences tribe" – and, perhaps, that these Geology students had not had the benefit of an Oberlin-like liberal arts education.

At the beginning of Bill's third year at Harvard (1965) and now working on his doctoral thesis on microscopic fossils of the 800-million-year-old Bitter Springs Formation of central Australia, he was elected as a junior fellow in the Harvard Society of Fellows – an amazingly great honor and opportunity. Established in 1933 under the terms of a gift from Harvard President Emeritus A. Lawrence Lowell, Bill's selection meant that for the next four years he would not be subject to examinations, would have no course requirements, and would receive a whopping annual stipend – most recently, in 2017, $74,000 – just a tad less than that of an incoming Harvard assistant professor.

Every Monday evening during term-time, the 24 junior fellows met for dinner with the Society's senior fellows, all notable luminaries – among them in Bill's day were Herbert Bloch (Ancient History, Classical Philology, and Archaeology); Paul Doty (Chemistry and Biochemistry); Willard Van Orman Quine (Philosophy and Mathematical Logic); Charles E. Wyzanski (Law); and Nobel laureates Edward M. Purcell (Physics) and James D. Watson (Genetics and Biochemistry).

When Bill first met senior fellow Van Orman Quine, so as not to embarrass himself, he tried to come up with some seemingly neutral question that would not reveal too much of his ignorance. Bill asked: "With whom, Sir, did you study?" Quine's response: "Alfred North Whitehead." Bill was floored! In his freshman philosophy courses, writings by Whitehead directly followed those of Aristotle and Socrates! And because senior fellow and Nobel laureate Jim Watson was housed in the Biology Building adjacent to Bill's, they walked together to dinner every Monday, during which Watson regaled Bill about his then-ongoing battle with Francis Crick (co-discoverer of the structure of DNA) who had threatened to sue if Watson's bestseller *Double Helix* were ever published in the United Kingdom. All this was a wonderful opportunity – *Bill was terrifically lucky, once again!*

As a graduate student, Bill gave talks at national scientific meetings and was author or co-author of nine publications (including his doctoral thesis, the first full-fledged monograph in the field). And after the landmark 1965 Barghoorn-Tyler paper on the Gunflint microbiota was published, the lecture requests to Professor Barghoorn became a bit more than he could handle, so Bill was often asked to lecture at various universities in his stead.

In the fall of 1967, the job market was rich with opportunities, and offers started to roll in (some 12 in all, including Harvard, MIT, Texas, Houston, Penn, Cornell, and the University of California Los Angeles [UCLA]). To Bill this was heady stuff and simply too good an opportunity to pass up. He decided to forgo his final year as a junior fellow and venture into the real academic world. Barghoorn wanted Bill to stay at Harvard. Frank Press, then head of the Department of Earth and Planetary Sciences at MIT (and later, science advisor to US President Jimmy Carter and president of the National Academy of Sciences), wanted Bill to join his department, telling him that MIT would "meet or beat whatever Harvard is offering." During Bill's job interview visit, however, as Professor Press showed Bill his lab-space, it became clear to Bill that none of Press' faculty colleagues were even remotely aware of Bill's possible appointment. Frank Press deserves much credit for having revitalized the previously "traditional" MIT department, but – given Bill's values of openness, honesty, and respect for others – Press' exercise of concentrated authority was concerning. Moreover, and most importantly, Bill wanted to be on his own, not at Harvard and not down the road at MIT. He selected UCLA, which has been wonderful.

5.3 UNIVERSITY OF CALIFORNIA, LOS ANGELES

Forgoing the graduation ceremonies at Harvard, Bill headed off to California. On the cross-country journey to UCLA, he stopped in Columbus to see his parents, during which visit he asked his father for advice as to how he should proceed in the professoriate. His father thought for a while and then suggested: "Make yourself indispensable," sage advice that over the years Bill has repeated to each of his students at the completion of their degree program. His father then paused again for quite a while, apparently wondering whether he really wanted to reveal to Bill his other advice, and then said: "*Be kind to the students.*" Bill's father then went on to explain that during his final semester at the University of Wyoming, he had been enrolled in a paleontology course taught by Samuel H. Knight (renowned as "Mr. Geology of Wyoming") and that during the final exam week, he saw professor Knight in the hallway and inquired, "Sir, when is our final exam?" Knight replied: "Jim, it was two days ago." Bill's father had received an "A" on the mid-term exam and now an "F" on the final. Professor Knight gave him the "C" in his course that enabled him to graduate on time, an act that Bill's father said to be "the kindest thing that had ever happened to me."

In June 1968, a newly minted Ph.D., Bill joined the faculty at UCLA where he has remained for 50 years (and 50% of the university's existence). Soon after his arrival, he headed off to carry out field work in Australia on the first of his many visits to the "Land Down Under," an excellent experience during which he met all of the (then few) Australian scientists interested in Precambrian Paleobiology, most notably Martin F. Glaessner, famed for his studies of the very latest Precambrian (560-million-year-old) Ediacaran Fauna of South Australia that contains the earliest known (soft-bodied) animals. And during this visit Bill also became acquainted with Glaessner's star student, Malcolm R. Walter.

Upon his return to UCLA, Bill was greeted with open arms by his department colleagues, who have been enormously kind and supportive ever since. Two years after Bill joined the faculty, he was promoted to the tenured position of associate professor, and that same year, 1970, his son was born, James Christopher (named after Bill's father and his wife's younger brother, Christopher Morgan). Bill cried with tears of absolute joy! Then, in 1973, after only five years at UCLA, Bill was promoted to full professor at the age of 31 (alleged to be the youngest in the history of the department) – and over subsequent years, he has been awarded all three of the UCLA campus-wide accolades available to members of the faculty: for teaching, for research, and for academic excellence. *Lucky, again and again!*

5.4 UCLA DEPARTMENT OF GEOLOGY

When Bill joined the UCLA faculty, Geology was a small department composed of 12, maybe 14, members. In retrospect, this small size carried huge benefits. The department was a close-knit "band of brothers" (entirely male, women at the time being dissuaded from going into this "physically demanding" field of science and excluded from entering underground mines – as they almost universally still are). Each month, one or another of the faculty would host a pot-luck dinner at his home where the faculty and spouses would get to know one another's families, including their children, and play group games – "Charades" being the group favorite. For all of them, these get-togethers were truly great fun!

Moreover, faculty members helped one another. Bill was teaching four, sometimes five courses per year (and thought that he was being misused, having hardly any time for his research – even though, as always, then and now, he was in the lab seven days a week). After a year or so, Bill was assigned as the third of a trio (Gary Lane, Clarence Hall, and him) teaching a lower division course on the History of Life, with Bill handling the paleobotany. This was a real boon! Gary Lane (also an Oberlin grad) taught Bill how to teach, his mantra being "the students come first," an attitude that Bill has ever since, at all times and in all ways, tried to follow. (Heck, he told himself, whatever extra time it takes him to help students he can make up by working Saturdays and Sundays.) And Bill finally knew that he had become fully accepted by his senior colleagues when, following Clarence's meticulous directions, Bill was permitted to prepare packs of celery sticks for their class field trip to Nevada! A few years later (and without Bill knowing), Gary and Clarence cajoled the class into supporting Bill's nomination for the university-wide Distinguished Teaching Award (which Bill received in 1977, the first in the history of his department).

One other little tale from this early period at UCLA deserves recounting. Bill was young, not much older than some of the department graduate students with whom he got along well, and Bill liked his faculty colleagues as well. But the students and faculty didn't generally interact on a personal basis. Bill thought that the situation could be improved, his solution being to set in place a late Friday afternoon gathering where the students and faculty would come together, as a whole, over pizza and beer, for which Bill coined the name after the geological-geochemical term "Liquidus." This was in the early 1970s when alcoholic beverages were prohibited on campus, even at the then newly forming UCLA Faculty Center. So, Bill invited the leaders of his department's student organization into his office, explained the situation, and left it in their hands – Bill's hope being, as he explained to them, that if the students behaved themselves properly, they could use this opportunity to invite the faculty to join them in this work week–ending informal get-together. That worked wonderfully – students inviting faculty to join the throng – but only for about six weeks. Liquidus then evolved into a mostly student-dominated week-ending group party as it has continued to this day. To the best of Bill's knowledge, the students have always behaved properly, despite the convivial atmosphere and the availability of beer (which, for all Bill knows, may still be outlawed on campus). *Hooray for them!*

After a few years, Geology merged with an even smaller department, Geophysics and Astrophysics (in shorthand, "GASP"). Since then, the department has had various name changes: from "Geology," to "Geology and Geophysics," to "Earth and Space Sciences," to its current moniker, "Earth, Planetary, and Space Sciences." Over time, the science has evolved and the department has too. Great! But as the department expanded, the group became too large for the monthly pot-luck get-togethers, and the close friendships, camaraderie, mutual support, and first-hand knowledge of one another's science – and of their families – slowly became a thing of the past.

Over the years, Bill has tried to offset this trend, to him unforeseen and unwanted. In that effort, for an extended period (12, perhaps 15 years) Bill and his wife hosted a yearly department party at their home. All seemed to greatly enjoy these gatherings – especially those attended by UCLA Chancellor Chuck Young or basketball coach Steve Lavin – but they ceased hosting these get-togethers when a new department chair was appointed who told them that he and his wife would prefer to handle this task. They never did. Not once. And it has never again occurred.

5.5 SOCIETAL CHANGES

The department, and the society, continued to evolve. Over the years women joined the department ranks as the number of women recipients of Ph.D.s in US Geoscience Programs rose steadily by 1% per year, from 10% in 1975 to the present nearly 50%. Great!

And attitudes changed also. Spurred by younger hires, the former department's "group-first" ethos evolved into a "me first" point of view (a "good business practice," one of the younger hires informed Bill), and previously accepted norms were increasingly viewed as inconsistent with "political correctness" (PC). On a few occasions, heartfelt but "over the top" unhelpful comments were voiced at departmental faculty meetings – largely but not exclusively by the younger cohort – and this "us" vs. "them," younger faculty vs. "old guard" departmental division was soon exacerbated by recurring episodes of the two-body problem (or, as some would prefer, "two-body opportunity"): the presence of two capable job-seeking Ph.D.s in the same family – one hired, the second seeking a position in the department.

Initially this problem/opportunity was a relatively new development in academia – an understandable if unforeseen result of the influx of women Ph.D.s into the Geosciences and their quite rational preference for similarly accomplished spouses. The department suffered through three such episodes in a scant six years. The three faculty members pushing the issue, all women and all well respected scholars, were quite naturally disgruntled (to put it mildly) when the department faculty, after careful thorough consideration, voted not to hire their spouses. After discussing the matter with the department chair (who no doubt did his best to offer solace) they and their cohort of young (in PC parlance, "early career") colleagues are reported to have journeyed over to the Administration Building (in Bill's parlance, the "Big House") and presented their grievances to the higher-ups in the UCLA administrative hierarchy.

Bill is by no means averse to activists. In fact, he generally applauds their efforts – he is, after all, a product of Oberlin College! So, while to Bill these episodes may have been a bit unsettling, they have neither been particularly surprising nor necessarily wrong-headed. The society evolves, the science evolves, and the proper role both of the university and the department need to evolve also, as they must, to adapt and adjust to the changing times.

The cardinal point is that although UCLA is ranked in the top ten among all of the so-called "elite universities of the world," its goal is to continuously improve – "Good Enough" is *not* good enough – to Bill's mind, one of UCLA's truly greatest assets. But openings for new hires are few and far between, averaging for a department of his size perhaps one, at most two per year. His department recruits faculty internationally and attempts to appoint only the very best – potential leaders of their fields – and, then, it offers appointments only in those in areas needed to fill its academic (research and teaching) mission. It is a great privilege to be a member of the UCLA faculty, the appointee having not uncommonly been selected from a pool of well over 150 applicants (i.e., the odds of selection being appreciably less than one percent). Given this, it would be unusual for two such highly gifted potential appointees to have become married or partners and rarer still for the un-hired spouse to be a prospective world-class expert in a field that the department perceived as critical to its future.

As a long-time fully committed member of his department, Bill was concerned. The very fabric of his academic "family," its collegiality and group-first ethos, were being torn apart. Thinking through the matter, Bill concluded that the basic problem was that the divided factions simply didn't "know" each other. That certainly was true of him. By then he had only very rarely met the spouses of his younger colleagues, didn't know their children, didn't know where they lived, didn't know their hopes and aspirations, didn't know how best he might try to help them. Perhaps for them, the professoriate is a "job" rather than being a "calling" as it is for Bill and many of his ilk. And Bill wasn't alone, his impression being that few of the department's "old guard" had personal knowledge of their younger colleagues. And if the "old guard" didn't know about them, they could not know about their senior colleagues either. But because of this – their notion of being an up-to-date "right-minded minority" stymied by an out-of-date "old school" majority – they made a huge mistake about Bill and the department. Not knowing of Bill's values or his

Oberlin background, they assumed that he was simply an "old white male," not to be trusted and neither interested in nor empathetic with their plight. In fact, one even referred Bill to an article highlighting the evils of racism, not realizing that Bill has battled against racial basis all his life.

Interestingly, and perhaps oddly, the divisive attitude that was pervading Bill's department strikes him as a bit like what he perceives to be the current (2018) ongoing stalemate-resulting difficulties in the US Congress – a "my way or the highway" factionalism that defeats progress for the society as a whole. In its stead, it seems to Bill that the way forward is for such groups to operate more like microbial symbiotic consortia, in which members of disparate biological lineages survive and indeed thrive as a result of their interactions with others in the ecosystem.

This analogy between cooperative microbes and preferred interactions in US Congress – which Bill particularly likes – stems from his correspondence with Judge David Tatel, a long-time member of the Washington, D.C., Federal Court of Appeals and a fellow member of the American Philosophical Society who has an interest Bill's work.

A couple of years ago when Bill discovered the first such consortium recorded in the fossil record (a 3,400-million-year-old assemblage of sulfate-reducing bacteria that produced foul-smelling hydrogen sulfide gas used to power co-existing photosynthetic bacteria), Judge Tatel suggested that Bill write an op-ed piece about his find and the analogy for the *New York Times*. Bill might have been able to sell the case – after all, *if literally billions-of-years-old microbes solved the problem, couldn't we humans, too?!* But Bill chose not to follow up on Judge Tatel's suggestion, largely because Bill imagined that the *Times* readership would not have found it particularly enlightening – it being yet another confusingly irrelevant example proffered by some card-carrying member of the "academic elite" – not to mention the umbrage some might take by Bill's comparison of billions-of-years-old dead fossil bacteria to elected members of Congress! And, anyway, Bill does not wish to be regarded as some self-promoting "know-it-all" seeking notoriety.

As the cooperative fabric of his department, Bill's academic home for so many, many years, seemed to be crumbling, he wondered: "What can I do?" In answer, in 2015, his wife Jane and he set up two endowments to the department, one to provide funds in perpetuity for a lunch to precede the department's monthly faculty meeting (meant to promote intra-faculty one-on-one camaraderie) and fund an annual faculty-staff dinner (when they hoped that the department staff would be publicly honored for their exceptional contributions). Their other endowment is to fund an experimental program intended to begin to ameliorate the academy-wide "two-body problem" by providing monies to employ an independent firm to identify and arrange contacts with potential employers in the L.A. Basin for the un-hired member of the pair.

As of this writing, the monthly luncheons and annual dinner have already proven beneficial. But the success of their first-step solution to the UCLA and University of California system-wide "two-body problem" is likely to take a good deal longer – 10, 15, perhaps 20 years. So be it. Some years ago, when Bill was a member of the Oberlin College Board of Trustees, he discovered that the small colleges of northern Ohio had already addressed this problem, evidently quite effectively. UCLA and the rest of academia can too. Over time, they will learn. As Chinese Chairman Mao Tse-tung famously said, "A journey of a thousand miles starts with a single step" (a phrase recast from that of the Chinese philosopher Lau Tzu, 604–531 BC, a contemporary of Confucius: "The journey of a thousand miles begins beneath one's feet").

6 The 1970s – Lucky, Time and Time Again

6.1 NASA'S LUNAR SAMPLE PRELIMINARY EXAMINATION TEAM: 1969–1971

A few months after Bill's arrival at the University of California, Los Angeles (UCLA), he was appointed to NASA's six-person Lunar Sample Preliminary Examination Team (LSPET), on which he served from 1969 to 1971, spending several months each year at the Johnson Space Center in Houston, Texas, where the team prepped for the Apollo 11 and 12 missions and then carried out the initial analyses on the lunar rocks they obtained. He very much appreciated UCLA's permission for him to be involved in this venture – and was also pleased, if repeatedly a bit embarrassed, by being called "Moon Man" by department colleague Ken Watson and his wife Mary, both of whom delighted in using this epithet.

For Bill, a fledgling academic, participation in LSPET was heady stuff, a real treat! The other members of the team were all senior established experts – Ed Chao and Bob Smith, highly experienced USGS geologists/geochemists; Professor Cliff Frondel from Harvard, the acknowledged "Dean" of American mineralogy; Marty Favaro from the US Public Health Service Centers for Disease Control in Atlanta, on-hand in case infectious agents were detected in the Moon rocks; and Jim Menzies of the US Department of Agriculture, the official approval of which was required to allow entry of "soil and rocks from a foreign entity." They all became great friends.

Prior to the July 1969 arrival of the Apollo 11 samples, the LSPET team had a couple of month-long practice runs, learning the "ins and outs" of the Lunar Receiving Laboratory (LRL) and developing appropriate protocols for handling and processing the soon-to-arrive samples. The LRL was divided into two sections separated by a sealed partition that in an emergency could be breached: A smaller crew reception area and a larger sample-processing facility. The crew reception area was designed to house the astronauts upon their return and, during their subsequent quarantine, their attending physicians. The sample-processing facility consisted of one huge room that housed an extensive interconnected multi-compartment vacuum system (in which the rock boxes were to be unpacked and their contents sorted under lunar-like near-vacuum conditions) and several adjacent analysis labs. Each analysis lab was equipped with nitrogen-filled glove box-type fume hoods operated at negative pressure (so that were the gloves or hoods to develop leaks, air would rush inward, not outward, protecting the workers and the lab environment from any Moon-derived pathogens). Near the entrance to the LRL was a medium-sized (150-seat) auditorium in which the two sections of the building were separated by a thick glass partition, a briefing room that would separate the quarantined astronauts from the media during press conferences and in which the team and other NASA groups could quiz the astronauts on details regarding the collected cache of lunar samples.

During these practice sessions they learned that access to the processing facility was to be strictly controlled (sign-in, sign-out); closely monitored (via closed circuit TV); and would require "showering-in," "showering-out," and passage in the nude through a sanitizing UV-flooded corridor (eyes to the floor). For those on the team wearing prescription eye-glasses, duplicates were to be provided inside (Figure 6.1). (Later, when the NASA authorities discovered that LSPET member Bob Smith had a hearing-aid that could not be readily duplicated, they issued an "Official Waiver" – a process that occurred repeatedly throughout the Apollo 11 and 12 missions.) And, they were to work eight-hour shifts – mostly during the "day shift" but sometimes three of the team during the day and the other three at night – with the third shift from midnight to 8am being set aside for "any necessary repair of facilities."

Other items of note occurred during the pre-mission sessions. (1) A plan afoot to test the "real" containment provided by the sample-processing fume hoods by seeding them with anthrax – ostensibly while LSPET was practicing protocols! – was thankfully scuttled. (2) In case of emergency – fire or, more fearfully, an outbreak of a Moon-derived infectious agent (which LSPET privately dubbed "Gorgo") – the team was to assemble on a lawn outside the building to be helicoptered to a nearby military base and from there to be airlifted to Bikini Atoll in the Pacific (site of the 1940s–1950s nuclear bomb tests) where it would remain ensconced until the problem was sorted out. (3) Analyses of the possible presence of infectious agents – of particular scientific interest to Bill – were to be carried out by an outside contractor,

FIGURE 6.1 The author as a LSPET member in the Lunar Receiving Laboratory (LRL).

Brown and Root-Northrop, rather than by the LSPET team or NASA employees (such arrangements even then having been an established NASA procedure to augment its staff when necessary and presumably thereby circumvent congressionally imposed hiring constraints). And (4) Bill learned that he was to be paid some $200 per day for his participation in the missions, whereas the four government-employed "reassigned" LSPET colleagues were not. To Bill, it was unfair for him to receive such largesse whereas these senior scientists did not. He wrote a formal letter expressing his displeasure and suggesting that NASA not pay him anything extra, either. The letter worked its way through the hierarchy and he was then informed: "That is the law." And so it evidently was.

In mid-1969, the Apollo 11 Command Capsule, with astronauts Armstrong, Aldrin, and Collins and the lunar rocks aboard, returned to Earth. The LSPET team again headed off to Houston and watched as the rock boxes traversed the extensive vacuum chamber to end up where they were to be opened by the stocky muscular glove-box operator. The boxes were already open, having never been sealed on the lunar surface, so this was a piece of cake! The team stood in line to have a look. Bill was second in line and, thus, the second of us on the planet to view rocks collected on the Moon. (They were dark gray to black, covered in dust – looking much like rock samples collected from a disused dusty coalfield; Figure 6.2.)

When the samples came to the analysis labs, LSPET began its work, but various problems soon arose – none, apparently, revealed to the public. During the Apollo 11 studies, the vacuum system of the rock-receiving chamber continuously failed, once or even two times a week. The midnight repair crew had their hands full, and the inability of the LRL managers to produce a quick fix eventually led to replacement of the facility director, Professor Oliver Schaeffer of the State University of New York, Stony Brook, one of the world's leading experts on large-scale vacuum systems.

Later, a Japanese quail – one of the test animals being "challenged" (i.e., being fed lunar dust to monitor its effect) – pecked its head though a rubber glove of its containment hood. When this "spill" (a breach of containment) was discovered the following morning, the emergency siren blared forth, and those in the various labs had to cease work and remain in place for the following several hours until an

FIGURE 6.2 Apollo 11 Moon rocks in the LRL vacuum chamber.

"Official Waiver" had been issued. And the test colony of sterile mice (delivered by Cesarean section and never exposed to infectious agents) rapidly began to die at an alarming rate of some 10% –15% per week – though it may actually have been even higher. Although members of the LSPET "lunar rock team" were not permitted to have a look at the mouse colony, they had the impression that the apparatus in the containment hood in "mouse lab" was less than sterile or that the hood itself was peppered with pin-hole leaks – or, perhaps, that infectious agents were flooding into the chamber when food was delivered to the struggling mouselets. In any case, germs to which the mice had no previously acquired immunity were decimating the colony. Replacement mice were being regularly flown in, but the outside world was never informed, NASA's official statement being (quite accurately) that "the mouse colony remains in its normal status."

Then, a potentially major problem emerged. The toilets in the sample-processing facility began gurgle, bubble, and fill to overflowing. Initially, this seemed more of an inconvenience than a problem. To protect the outside environment, all waste from the building (solid waste, gaseous effluents, everything) was decontaminated and incinerated. The building, and thus the waste-processing system, had been designed to accommodate a relatively small number of scientists, other workers, and the astronauts and their physicians. However, soon after delivery of the Moon rocks, the building began to be flooded by visitors, primarily local elected officials and members of Congress. These visitors did not have to submit to the "shower-in, shower-out, UV-flooded corridor routine" – and they were not permitted in the lunar rock–containing science labs (though the scientists could see them peering in through the window in their doors to their labs) – but, quite reasonably, they were permitted to use the toilet facilities. The assumption of the LSPET workers and that of the LRL managers was that the waste-processing network had simply become overloaded.

Once the toilet malfunction was discovered, the siren sounded, the scientists stopped their work, and the lab managers got out the building plumbing plans. They soon realized that toilet facilities and the other plumbing in the two sections of the building were interconnected. This presented a serious problem. Because the astronauts had not been able to seal the rock boxes on the lunar surface (as was called for by the NASA protocol), on the return flight, the gravity-free Command Capsule became filled with floating lunar dust. To the LSPET team, this was not "new news" as they had seen videos of the crew during the return flight (when they looked like dust-coated coalminers) before they cleaned up for their national TV broadcast on the evening news programs. But the team also learned that the astronauts had dropped several of their film canisters on the lunar surface, and because the films were being processed in the crew reception facility, the dust from those canisters and that still clinging to the astronauts' space suits and to the astronauts themselves had all been flushed down the drain into the labyrinthine plumbing system. In principle, everyone in the interior of the building – on both sides of the separating partition – had been exposed to Moon-derived germs!

As Bill and his colleagues sat and waited in their lab for some 8–10 hours – deprived of bathroom breaks (a major problem for one of his colleagues, which he solved by urinating into an Erlenmeyer flask) – the powers-that-be ultimately decided that there was simply nothing to be done. Another "Official Waiver" was signed, and after a completely wasted day, the team was finally permitted to depart and return to their hotel rooms.

Other interesting items similarly never received public notice. For example, if you watched the news conferences for Apollo 11 and 12 while the astronauts were in quarantine, you may have wondered why they always wore baseball caps emblazoned with their mission's logo. In actuality this was not part of their uniforms and not just for show. In those days, the Command Capsule was filled with an atmosphere of 100% oxygen, and the skin of each of the astronauts was wired with numerous sensors to continuously check their temperature, heart-rate, blood flow, and so on. Such conditions are ideal for the rapid growth of the common skin bacterium *Staphylococcus aureus* (known to hospital workers acquainted with oxygen tents as the "golden pus bacterium"), which produces pus-filled abscesses where the skin has become even minutely abraded. Upon their return the astronauts were pock-marked with scattered pustules, not just beneath where the sensors had been attached but also behind their ears and on the crowns of their heads where their head-sets had been held in place – unsightly lesions hidden from public view by the baseball caps and a condition rapidly remedied by antibiotics administered by the attending physicians during the two-week quarantine period. (The six LSPET members were relieved that the public didn't know about these infections, reasoning that if word got out, there would likely be widespread concern that the dreaded "Gorgo" was on the loose and that the bunch of them would end up marooned on Bikini Atoll!)

The first two sets of astronauts were a select, courageous, close-knit band, all but Neil Armstrong sharing a military background. They liked to goof around and play jokes on one another. (This was great for their camaraderie but not so much for lunar science, Apollo 12 astronaut Alan Bean being the only member of the group who seemed genuinely interested in the rocks collected and the science they would enable to be carried out.)

As you watched the televised interviews with the Apollo 12 Conrad–Gordon–Bean crew during their quarantine period, you may have wondered why they sometimes grinned when they became bored and sneaked furtive glances to the inside wall of the briefing room to their right (to the viewer's left). The apparent answer: During the Apollo 11 Armstrong–Aldrin–Collins crew's post-quarantine cross-country "Victory Lap" visit to San Diego, they had briefly visited Tijuana, Mexico, where they are said to have purchased a six-foot high painting on black velvet of a lithesome nude woman, which they hung off-camera on the inside of the briefing room to distract and amuse the follow-on crew. Though the Apollo 12 crew evidently never publicly mentioned this painting – and none of the LSPET team personally viewed it – the playful demeanor of the astronauts during the science debriefing sessions suggests that this story is likely to be true.

In sum, Bill's LSPET experience was remarkable, truly wonderful. And for him it began a decades-long interaction with NASA that was to become crucial to the establishment of the then-nascent field of Astrobiology, to the formation of NASA's highly successful Astrobiology Institute (the NAI), and to NASA's Precambrian Paleobiology–based strategy for the search for evidence of past life on other worlds.

6.2 TRAPPED IN THE INDIA–PAKISTAN WAR: 1971

Over the UCLA Christmas break of 1971, to attend a scientific meeting at the Birbal Sahni Institute of Paleobotany at Lucknow in north-central India, Bill flew to Delhi and then visited the Taj Mahal and the "Red Fort" at Agra. As his taxi returned to Delhi, he noted to the driver that the city was pitch black, no lights at all. The driver explained, "Sahib, don't worry about it. It's just another air-raid drill in case the Pakistanis attack." The next day, December 6, a copy of the *New Delhi Times* was shoved under Bill's hotel door. The headline, in three-inch-high bright-red capital letters: "WAR WITH PAKISTAN!!"

That morning, Bill took a taxi to the American Embassy and talked with three or four of the Marine guards (who were about his age): "*Don't be concerned*," they told him, "*we have an extra bed and you can move in with us – you'll be safe here.*" Still, Bill had to at least try to get to the Lucknow meeting, so he took a taxi to the airport to catch his plane. His flight was cancelled and he was placed on another – then another, and yet another. The morning wore on as the planes were repeatedly commandeered to transport troops. (Indeed, for the following two weeks, plane travel within India was completely disrupted and all foreign flights were cancelled.)

It appeared that Bill was to be stranded in Delhi, but a kind Indian gentleman (a corporate executive of Madras Rubber Factory Ltd., an India-wide tire and rubber company) took him under his wing and told him where to go to obtain a train ticket; how he should there bribe his way into a sleeping compartment for the all-night trip (a notion that in his naiveté would never have occurred to Bill); and how to deal with the porter at the Delhi train station. Following these welcome instructions, Bill made the arrangements and arrived that afternoon at the station at the pre-curfew hour of 5pm. He then located a porter whose number (#313) he jotted down in the porter's presence, to let the porter know that he had a firm record and, as Bill had been instructed, he paid the porter half, but only half of the required fee – to make certain that that the porter would actually return later that evening to guide him to his midnight train.

At the appointed hour, the porter returned and they trundled off, the porter striding ahead with Bill's suitcase balanced on his head. The station was completely devoid of light, pitch black like the innards of an enormous cave, but around them they could hear chickens and pigs (no cows) and great swarms of milling people muttering in Hindi. The porter understood no English, and Bill, no Hindi. But the porter knew where he was headed and Bill did not. As the porter continually charged off in front, Bill called out "313 – Stop!" and when Bill caught up, "313 – Go!" They reached the boarding platform and located Bill's compartment. But the porter had forgotten to obtain Bill's bedding, without which he would not receive the other half of his fee. Off he scurried to return in less than 10 minutes with the bedding (which, in the blackness of the station, Bill felt certain he had purloined from some unsuspecting traveler). Bill paid the remainder of the fee. All seemed well.

Bill was soon joined by two fellow travelers on their way to Lucknow, an elderly woman palynologist from Germany who was to attend the same meeting as was he, and a middle-aged, evidently quite wealthy, owner of numerous Delhi movie theaters.

The woman palynologist, terribly distraught, spent most of the next ten hours weeping, off and on throughout the night and into the next morning. The compartment was designed to hold four passengers on a pair of bunk beds, but after some 15 minutes of waiting, the fourth occupant had not arrived. The movie mogul declared "That's long enough. If you need to, visit the toilet now because I am going to seal the compartment." As the palynologist and Bill stayed put, he then hauled out three huge pincer locks, placed them at strategic positions on the inside the door and locked them in, explaining that he was transporting a huge sum of cash, the monthly proceeds from his Delhi movie theaters; that for safe-keeping he had previously given half to the train conductor (with the remainder in a thick money belt wrapped around his rather fulsome belly); and that his locks would keep out the marauders who would attempt to enter the compartment at each and every stop. (He was correct. The marauders did come. The locks held.)

Late in the next morning they arrived in Lucknow and departed the train. The German palynologist, now only sniffling from time to time, was relieved, as was Bill.

The scientific meeting then commenced. All windows at the meeting venue were covered with black-out curtains and all of the previously planned evening events (dances, shows, and celebratory dinners) had been cancelled. There were six non-Indian visitors (out of an original roster of some 30) who had now become trapped in the country, all international flights having been cancelled. Among these strandees was Aureal Cross, the paleobotanist at Michigan State University, accompanied by his wife Alene; and Dan Axelrod, "Ax," Bill's immediate predecessor at UCLA. There being so few foreigners on hand, it was paramount for at least three or four of them to be present for each and every scientific session, whether they were interested or not. Bill played his part.

During the meeting, the guests had one afternoon trip to the Lucknow Zoo, where Bill mounted a docile elephant, much to the delight of Aureal and Alene Cross (both of whom had known him since he was a baby and at whose wedding Bill's father had served as "Best Man"). And most nights, Ax and Bill sat in the hotel bar, where Ax regaled him with stories of his previous long tenure at UCLA as a member of Bill's department.

As the meeting continued, the short (two-week) 1971 India–Pakistan War raged on. The Indians were mobilized and fervent as they prolonged the conflict in order to "liberate" East Pakistan and thereby help establish the new country of Bangladesh. As Bill walked back to his hotel each afternoon after the day's session, he was repeatedly accosted by the locals who were irate that the United States was reported to have "tilted" toward Pakistan. To defuse these confrontations, Bill replied in German rather than English – a good ploy; they left him alone. After the meeting, Bill took a long train ride from Lucknow to Bombay (now Mumbai), caught the first plane permitted to leave the country, and returned home to California for Christmas.

On the face of it, this first India trip could not reasonably be termed a "lucky episode." But for Bill it actually was. He made contacts with Indian scientists that soon proved highly beneficial, paving the way for him to revisit India two years later to carry out highly rewarding field studies of the Precambrian strata of southeast India and learn a bit more about Indian culture. *He was fortunate, once again!*

6.3 FIELD WORK AND A GLIMPSE OF INDIAN CULTURE: 1973

Bill's host during his follow-up three-week trip to India was K.N. Prasad of the Geological Survey of India (GSI) Southern Region. Although Dr. Prasad specialized in strata of the Pleistocene Epoch (the most recent 2.6 million years, the great "Ice Ages" from which the ice sheets of Greenland, the Arctic, and the Antarctic still remain), he was well versed on the geology of the Precambrian Cuddapah Basin of south-central India, an extensive area in the vicinity of Hyderabad, the location of his home office. The GSI was welcoming to their American visitor and exceptionally generous, providing a fleet of three four-wheel-drive jeep "carryalls" (forerunners of current SUVs, all seemingly World War II vintage) and accompanying GSI staff: Three leaders (all geologically well-educated Brahmins), each of whom had a personal field assistant; and three drivers.

The first stop of the team was at the Brahmanapalli Asbestos Mine, near Hyderabad and some 150 miles northwest of Madras (now Chennai), where Bill collected stromatolitic limestones from the 1,400-million-year-old Vempalle Formation. After their work, they stopped at the mine headquarters to pay their respects. Because this was a Sunday afternoon, there was only one person in the main office – a middle-aged watchman/caretaker who was amusing himself by staging fights on his desktop among three large scorpions that he controlled by threads looped around their midsections. He then showed his visitors bottles containing the ten other scorpions that comprised his "fighting stable" and noted that on occasion he would get them all out at the same time and stage "a real battle."

Because Bill was officially a GSI visiting scientist, the group had to be housed each night at one or another "Government Guest House" – a well-intentioned but inconvenient arrangement that required long drives to and from their work areas. Each day they collected rock samples from various outcroppings, which introduced Bill to the ways of Indian geology. At the outcrops, usually at road-cuts where the rocks were well exposed, the three Brahmin geologists would each point to a particular rock up the slope from where they were standing and instruct their field assistants to collect them. Frequently the geologists would then glance at the returned sample for a few seconds, toss it away, and then point to another, sending the assistants scurrying back up the slope, a process that was repeated over and over. Bill hadn't seen this practice before, so he asked Dr. Prasad about it. Prasad's reply: "Oh, it's only good sense; it protects the geologists from scorpions that might hide under the rocks" (though, of course, the geologists' field assistants might have had a differing opinion).

Later that first week, the field-group's caravan went off-road onto a farm track used by ox-carts. For the first mile or so, all went well, and then the first vehicle abruptly stopped. Everyone got out to have a look at the problem. There at the left-center of the track was a large 12-inch-high chunk of granite protruding from the underlying rock mass, too high for the GSI jeeps to traverse without tearing out their underpinnings. There was no room to maneuver around it, either to the left, where the track was fronted by a steep bank of black slate, or to the right, where immediately alongside was a precipitous 30-foot drop. The track was fine for high-draft ox-carts but not for even four-wheel-drive jeeps.

While Bill was off toward the back of the group with the field assistants, the leaders huddled together at the front to sort things out. Dr. Prasad then walked over and told him, "You'll see we're prepared, we brought along our own road-repair crew." Almost immediately a tall, scrawny, emaciated-looking elderly man emerged from the back of the vehicle carrying the rock-hammers and the samples they had collected. Rather than being permitted to ride in a passenger seat, he had been relegated to the cargo compartment at the rear of the carryall, where he sat on the rock bags and hammers. Bill had never before seen this person – he didn't know that this worker was with them on the trip – but Bill assumed him to be a Shudra, a member of the "servant class," the lowest of the four classes or castes of Hinduism (a rank-ordered system officially abolished in 1950). The other possibility was that he was an "untouchable," exiled from the Hindu caste system – but that seemed unlikely, given that this was a GSI expedition.

With a bulky sledge-hammer in hand and having received his instructions in Hindi, the rest of the group stood off to the side and watched the old man attack the granite boulder in their path. Four, five, six times he struck, as hard as he could, but to no avail – not a single chip went flying away. The blows came slower and slower as he panted and wheezed. He seemed spent, nearly exhausted. Bill was concerned. This plan was not working. His first instinct, given his background, was to get in there and help the old man – maybe the two of them, together, could defeat this problem. But Bill knew his place. He was merely a guest (not an Indian, not a Brahmin) – it was their culture/society, not his. Bill's proper place was to remember the old adage: "When in Rome (in this instance, India) do as the Romans (i.e., Indians) do."

Still, as Bill thought through the problem, he came up with a solution. The slate bed to the left of the road was solid, geologically thoroughly pressure-cooked (metamorphosed). His idea was to take out a bunch of these strong slate slabs, lean them against the right side of the granite boulder, and, slightly tilted, drive the left wheels of the vehicles over this supporting ramp. But it was not his place to suggest this to the trip leaders – he was a visitor, an outsider. So, he suggested his notion to Dr. Prasad, who then returned to the huddling leaders and suggested it to them. They agreed. The slabs were set in place. It worked. And Dr. Prasad was credited with figuring out the solution. *Excellent!*

A week later, the group stopped again at the asbestos mine to collect more samples of the Vempalle Formation, which, to Bill, looked especially promising (and in which, after his return to UCLA, he discovered the first Precambrian fossils in India and the oldest of the Indian Subcontinent – a find that initiated the now vibrant studies of Precambrian Paleobiology across India). At the conclusion of this visit, the group again dropped by at the mine headquarters, where Bill inquired about the watchman they had previously met and was told that "He died last week from scorpion stings." As it turned out, the watchman's "fighting stable" included eight Indian red scorpions (known also as the eastern Indian scorpion, *Hottentotta tamulus*), especially large scorpions that are two to three inches long and reputed to be the most lethal on Earth. One can only surmise that for his amusement he had staged, as he had mentioned earlier, "a real battle" – with him being the thread-holding puppeteer of his 13-scorpion "fighting stable" – and that through his own careless play he had met his demise.

On a Saturday evening near the end of that visit, Dr. Prasad invited Bill to his home in Hyderabad where he met Dr. Prasad's wife and their son, Vinod. During that dinner, Vinod – enormously proud of his father – told Bill that as a journalism student at the University of Hyderabad he wanted to write a feature article for the Sunday Supplement of *The Times of India* to highlight his father's accomplishment of having designed and directed the construction two life-sized fiberglass reconstructions of dinosaurs (*Triceratops* and *Tyrannosaurus*) on display at the Hyderabad Zoo (formally, the Nehru Zoological Park) – where better to house dinosaurs!? Indeed, as Vinod recounted, his father's dinosaurs were so noteworthy that they had been filmed from helicopter by the British Broadcasting Company, the BBC. Vinod's problem was that he had no color film (the Sunday Supplement requiring such images) because color film was not generally available in India in 1973 and impossible for him to obtain. Bill, however, had quite a lot of color film with him in order to record his field work, and he was glad to help the young student.

The next morning, Dr. Prasad and Vinod fetched Bill from his Hyderabad hotel, and the three of them traveled out to the zoo so that Bill could take pictures of Dr. Prasad's creations. Positioned near the middle of a marshy area in which the life-sized reconstructions were surrounded by grasses, reeds and trees, the dinosaurs were excellent, spectacular! Their enclosure was marked by a two-foot-high white-painted metal fence, which Bill straddled over; he then walked off into the swampy area to get a close-up look at the dinosaurs. He shot quite a number of photos (Figure 6.3) and then, as he changed to a new roll of film, he glanced back at Vinod and his father. There beside them were about 50 on-lookers. Bill changed the film, took more photos, and then looked again to the crowd as he slogged back through the marsh toward the fence. The curious assembled on-lookers had now increased to some 300. Bill again straddled the small fence and the crowd began to furiously applaud and loudly cheer. Bill was embarrassed and perplexed. He asked Dr. Prasad: "Why all this uproar?"

FIGURE 6.3 Fiber-glass replica of *Triceratops* at the Hyderabad Zoo.

Prasad's answer: "Didn't you read the sign on the fence?" Bill hadn't noticed the sign, and it wouldn't have mattered anyway because, as he told Dr. Prasad, "No, I do not read Hindi." Dr. Prasad then translated it: "Stay … Out … Of … This… Area!! … There … Are … Many … Cobra … Snakes … Here!"

Before heading to Bombay (now known as Mumbai) to board his flight home, Bill journeyed north to Allahabad where he had been invited to give a series of lectures at the university. To Hindus (some 80% of the Indian population), Allahabad is a sacred city, primarily because it is situated at the confluence of the two most holy rivers of India, the Ganges and the Yamuna (known in 1973 as the Jumna) – the site, for example, in 1948, of the funeral pyre of Mahatma Gandhi, the assassinated founder of modern India and "Father of the Nation." Bill had seen a similar much-revered locale before in Varanasi (known also as Benares), where funeral pyres lined the bank of the Ganges and the ashes of the deceased were ritualistically sprinkled into the river; white linen-enshrouded bloated bodies floated by (placed in the Ganges upstream of Varanasi by family members too poor to afford the cost of cremation); and sacred cows wandered about in the shallow waters.

After one of Bill's lectures at the University of Allahabad, a highly knowledgeable microbiology professor invited him to his lab where he beamed as he told him that he had "solved the problem of the origin of life." Bill, of course, was immensely interested – what had the microbiologist discovered? The professor pointed to two conical (Erlenmeyer) glass flasks next to a window in his lab, the water in each of which was topped by green scum (which Bill imagined to be cyanobacteria). The professor then explained: He had personally collected the waters in these flasks at the exact confluence of the Ganges and the Yamuna – a site so holy that that it was "known to all to be absolutely sterile," devoid of any form of life – yet on the window-shelf in his lab, these waters had "come alive."

"Sterile river waters" with cows wandering around and enshrouded bodies floating by? This made no sense. But Bill listened, he learned, and he thought. What he learned from this and other similar episodes from his travels over the years is that all peoples across the globe are influenced by the shared history, beliefs, and cultures of their societies. Despite what his aunts in Iowa and New Mexico may have assumed, the Christian-dominated "American way" is not the *only* way to wend one's path through life!

Since these first two trips, Bill has traveled to India four or five more times, during which he visited the homes and met the families of numerous Indian scientists, all Brahmins. Not uncommonly, they "let him into their lives," explaining Hinduism to him and telling him how it is the glue that bonds their families and their society. As best Bill can tell, all have a "shrine" in their homes, typically a large closet with a small statue as the centerpiece, where the family gathers for an hour each day to worship, chanting and singing as strings of colored lights blink on and off. Because there are numerous deities in Hinduism, the household statue varies family to family, house to house … Ganesha, Brahma, Vishnu, Krishna, and others.

When Bill first encountered this practice – him being neither a Hindu nor a theist – he imagined to himself, "Wow, what a terrible waste of time!" However, as he later thought through the matter, he came to realize its value as a family-unifying ritual, not unlike the daily family dinners that his brother and he endured day after day when they had to report to their parents "what they had learned" that day. To an American, it may seem odd for one class of a society (in India, the Brahmins) to lord over another group (members of the "lower castes") and require the underclass to do the menial tasks and manual work. But in India that is standard practice, a societal norm that credits the assumed status-accompanying knowledge and experience of the educated Brahmins.

Interestingly, such status-related assumptions differ not so much from some well-accepted practices in Western societies. Consider, for example, Germany, where the title "Doctor Professor" is an official formal part of the bearer's name, required on driver's licenses and legal documents – a practice instituted a century ago to acknowledge the earned excellence of such title-bearers and raise them to equality with the European "*Van* so-and-so" and "*Von* such-and-such" for which their honorary prefix reflects their forebear's position in the established nobility. For such practices in India – or Germany, or elsewhere – one should remember that such norms are a product of their history, beliefs, and culture.

In 1949, when Bill's family moved from Pittsburg to Columbus, he witnessed this preference for the "educated class" first-hand. His parents needed a bank loan to purchase their new home. The loan was approved – at a rate appreciably lower than that of other borrowers – because Bill's dad had been appointed a professor at the Ohio State University. High-school and probably grade-school teachers as well were granted a similar, if somewhat less munificent, discount. Such practices, of course, date from the immediate post-World War II period, when educators were valued by society – a special "break" that vanished long ago. But the lesson remains clear. Though on paper, at least, America strives to be equalitarian, it is only fair to remember that it also sometimes misses the mark.

Recognition of culturally or universally shared biases and predilections is important to science: *What should one accept, how well are the relevant "facts" really known, and does it matter from whom and from what society those facts emerged?* In science, such biases are a prime factor to guard against, for uncritical acceptance of that which is said to be "known," the currently accepted dogma, is anathema to effective rational inquiry – as it was globally for a full century with regard to the vexing problem posed by Darwin of the so-called "missing" fossil record of early life, when the accepted dogma dictated that life's earliest history was both unknown and assumed to be unknowable.

Nevertheless – in modern human history dating from the Age of Enlightenment – it would be wrong to be overly concerned about the advance of science. Like the societies in which it is embedded, science may progress regrettably slowly, stumbling along in fits and starts, but the small advances ultimately add up. And, there being only one reality, firmly established scientific facts will ultimately always prevail. Yet, again like human societies, it is also a huge mistake to ignore the unquestionable truth that, just as the past has for science determined its present, it is that present – where it finds itself today – that will determine its future. In short, science is a product of the society and culture in which it develops. Much can be learned from an examination of their intermeshed interdependent histories.

6.4 ADVENTURES IN THE USSR – THE COLD WAR AND THE KGB: 1972–1975

In the early 1970s, Bill was invited to attend the founding meeting of ISSOL, the International Society for the Study of the Origin of Life (of which he was later to serve as president) in Pont-à-Mousson, France. He declined – he had classes to teach and thought it important to follow the dictum that "the students come first." Moreover, he knew full well that, like his father and other American scientists, sooner or later he would get to Europe. That seemed certain to happen, and in any case, he had other fish to fry. By that time, he had become a committed interdisciplinary internationalist and he wanted to spread the "gospel" of Precambrian Paleobiology – as he has throughout his career. He also had a fairly good understanding of the distribution of Precambrian deposits worldwide, and those of India, China, and the USSR, particularly its extensive Russian and Siberian Platforms, seemed ripe for harvest.

So, in 1972 and again in 1974, Bill was delighted to be invited to give lectures in Moscow. In those days – as the Cold War surged on – interaction between the two world powers had essentially ceased, the only line of non-governmental communication being that doggedly maintained by the two countries' National Academies of Sciences. And even these bodies were markedly affected, their exchange of visiting scientists having been steadily cut over preceding years from 60 to 48 and, by that time, to only 24 "man-months" per year. The United States used its 24-month allotment for 12 one-month visits and two "long-term" six-month sojourns.

Spurred by his previous two visits, Bill applied for and was selected as a 1975 "long-term Exchange Scientist" between the United States and the USSR Academies of Sciences. This, for him, was a life-changer. And, thanks to his previous visits, he had a notion as to what to expect and how to prepare. He enrolled in a one-on-one six-month Russian language Berlitz class (during which he brushed up on the rudiments of Russian he had learned during his 1972 and 1974 visits and then sprinted through the first two years of standard college-level Russian). And he planned the trip carefully. So as not be stranded in any single non-productive locale too long, he organized stays of two months each in Moscow, then Akademgorodok (near Novosibirsk, Siberia), and the final two in Leningrad.

His wife Julie, their four-year-old son Jamie and he arrived in Moscow at the beginning of January, 1975. The other long-term exchange scientist had an apartment down the hall. He was a mathematician from Stanford and had not bothered with language lessons because, as he told Bill, "mathematics is an international language." As he had apparently surmised, upon visiting his Soviet Academy Institute, he discovered that no one spoke English. However, to his surprise and chagrin, he also discovered that "no one was interested in his work." The following week, 10 days after his arrival, he abruptly departed to return to the United States.

Within two or three days after his arrival in Moscow, Bill received a telephone call ordering him to meet an officer of the Soviet Academy – a huge burly guy, probably affiliated with the KGB, who tried his best to intimidate this American visitor as he towered over Bill, arms outstretched, and bellowed out: "You must not spy!!" "This is pure baloney," Bill thought, and he felt sorry for his erstwhile intimidator.

The USSR Academy officer was only doing his job – odd as that job seemed to be – and even given his height and girth he was a lousy play-actor. He didn't know this visitor. Bill didn't know him. Bill was in Soviet Russia simply to do science.

About eight days later, Bill received a second phone call, this one summoning him to the US Embassy to have a discussion with the science attaché. The sole scientist Bill knew in his Moscow institute who had a car, Misha Kritsky, delivered him to the embassy but dropped him off two blocks away so that his license plate would not be photographed at the embassy entrance by the KGB. Misha knew the ropes whereas Bill did not.

Bill checked in at the front desk and the attaché soon appeared and ushered him into the embassy innards, where he saw the most surprising set-up imaginable. What struck him first was that though this Stalin-period high-ceilinged building seemed unremarkably standard from its outside, its interior had been subdivided by an "erector-set" array of metal flooring and partitions such that it contained twice as many floors (and twice as many workers) as he had imagined. The two descended the stairs, one set after another, and finally ended up in a huge room that to Bill was like something out of *Star Wars*. It was unbelievable! The walls were coated with what looked to be thick layers of aluminum foil, and in the center was a lengthy wooden bridge connected to a large Plexiglas sphere that contained a desk, a phone, a few chairs, and some lights. The attaché led his visitor over the bridge and, once they were settled in the sphere, told him "Here we can talk, in private, without the Soviets listening in."

The attaché never actually asked Bill to "spy." Rather, he noted (correctly) that because of Bill's pre-planning he was scheduled to spend time at three Soviet Institutes and lecture at five others, and he asked only that Bill provide him with a list of those Soviet scientists with whom he had come in contact who were "friendly" to the United States. Bill replied that at the conclusion of his visit, he would be pleased to send the attaché the report required by the US Academy of Sciences intended to aid follow-on long-term exchange scientists (which, in fact, Bill later did). The attaché's rejoinder: "That's not exactly what I have in mind – I want names, fields, interests and attitudes of the 'friendlies.'"

The attaché had his agenda and Bill had his. Bill knew immediately that this was something that he could not and would not do. Had he acquiesced, every interaction that he had with fellow scientists and other Soviet citizens as well would be colored by this need to report his assessment of their political views. Though Bill was not then, nor is he now, adverse to the activities of the CIA, FBI, DOJ, and so on – and the science attaché was, he presumed, only doing what he perceived to be "his job" – as far as Bill was concerned, the intelligence agencies had their responsibilities and he had his. Bill gave no commitment. The Attaché and he parted (as Bill lodged this encounter in his memory banks).

Unlike the American intelligence agents – or so one might like to believe – the Russian KGB seemed to know everything that Bill did and all with whom he came into contact, repeatedly interviewing his Soviet friends, especially if they had invited him to their homes for dinner. And Bill was repeatedly quizzed about his outside activities by the Foreign Office of the various institutes he visited. This was, after all, in 1975, at the height of the Cold War, and National Academies of Sciences exchange scientist Bill Schopf was evidently the only long-term American visitor "on the loose" in the entire country.

During a visit to Soviet Russia a couple of years later, the KGB surveillance continued. A translator in Leningrad whom Bill had met during his earlier visits, a middle-aged lady named Zenaida, invited him to dinner at her home to see her new apartment and meet her mother. As she had instructed, Bill took a taxi to a location six blocks from her apartment building and later departed by the same route. Less than an hour after his departure, she received a phone call (at her unlisted number) ordering her to appear at the Leningrad KGB office within 30 minutes, where she was detained for questioning from 2am to 7am. Later that morning, before she was to translate Bill's lecture, at her request he visited the Leningrad American Consulate and obtained two cartons of Marlboro cigarettes and five Bic lighters, which he delivered to her. The following day she told Bill "*vse horosho*" (all is well). She had passed along these prizes to her interrogators, a practice known in Russian as "*nalevo*," literally, "on the left" – small bribes used to fix unpleasant situations.

For Bill, the six-month 1975 visit went well – he made many friends, discovered six new Precambrian microbiotas (the first three-dimensionally preserved communities of such fossils known in the USSR), and opened his field of Precambrian Paleobiology to Russian scientists (some of whom still remember him as a "founder" of this aspect of Russian science). Though the visit was no problem for him, it was to his family. Son Jamie had a difficult time adjusting to Soviet pre-school – quite understandably, given that he was exposed to a new teacher and new playmates every two months, none of whom understood English. And wife Julie, who spent her days as a "tourist" in a country in which at the time tourism was non-existent, felt isolated and alone (despite her previous training in the Russian language). After five months of this ordeal, Julie had enough – she and Jamie departed to the United States. Bill stuck it out.

When Bill returned home, he wrote his obligatory report to the US Academy of Sciences. But the initial episode with the US science attaché still stuck in his craw. So when he wrote a "thank you" letter delivered via diplomatic pouch to American ambassador Walter J. Stoessel – to express his gratitude for the kindness shown to him by officers at the embassy in Moscow and the consulate in Leningrad – he appended (an easily removable) two-page postscript outlining his discussion with the science attaché. To avoid unnecessary embarrassment to anyone involved, before sending this missive, he cleared it both with colleagues at UCLA and with the foreign secretary of the US National Academy.

Several months later, Bill learned from a Moscow-based correspondent for *U.S. News and World Report* whom he had met during his visit that the attaché had been summarily dismissed – present on a Friday then gone the following Monday. Looking for a good story, the reporter asked, "Do you know anything about this?" Bill replied: "No. It's news to me." And, truthfully, it was. However, as he later thought through the matter and checked his records, he realized that the timing of this event jibed, precisely, with what would have been the delivery date of his private letter to Ambassador Stoessel. Then and now, Bill has no knowledge whether his letter was or was not the impetus for the attaché's dismissal. Evidently, however, his letter raised eyebrows at the CIA, which, as he learned some eight or ten years later, had contacted several UCLA administrators to determine whether Bill harbored "suspect views." The UCLAers vouched for him, support for which he is of course enormously grateful. But he had nothing to hide; he had done nothing wrong. So it goes!

Four other vignettes from Bill's 1975 Cold War adventures in Soviet Russia deserve recounting.

First, it is not exactly true that for him "the visit was no problem." In fact, night after night for well more than half of that six-month visit to Soviet Russia, Bill recalls having the same recurrent nightmarish dream: He was in front of a firing squad; he could ask only a single question to save his life; his question of the squad was "Who among you can speak English?" He obviously had been worn to a thin nubbin by his inadequate command of the Russian language – a taxing deficiency given his scant vocabulary and being surrounded daily by non-English-speaking colleagues – and his innermost self was thus seeking someone to plead his case. The stress was palpable. Inwardly, he was afraid.

The second vignette: A prime goal of Soviet Communism was to transform Russia into a "workers' paradise" exemplified by its then recent cut-back to a six-day work-week. But Bill was in the USSR for a rather limited time to do science, and prohibiting him from working seven days a week was anathema – even though no one was allowed to be working alone on Sundays, the official day off. So, in Akademgorodok, he "beat the system." He found a colleague at his institute who was willing to accompany him into the building Sunday mornings (legitimizing Bill's presence and, in the accompanying scientist's eyes, being an exemplary gesture that would prove his worth to the all-powerful Communist Party). They both then formally signed in and went to Bill's lab where he would get to work and his colleague would read a novel. After half-an-hour or so, the accompanying colleague would then lower himself out of the lab's ground-floor window as Bill continued to work. Late in the afternoon, Bill would depart and sign out for them both when the building watchman was fast asleep.

A third anecdote: On follow-up lecture visit to Moscow a few months later, Bill was asked to be interviewed on Radio Moscow, the principal national news outlet at the time. Knowing of his recent six-month visit the reporter asked, "Which city do you most admire, Moscow or Leningrad?" Having some knowledge of Russian history and the centuries-old Moscow–Leningrad rivalry, set in place largely at the time of Tsar Peter the Great, Bill knew this to be a loaded question – a conundrum from which he escaped unscathed. His response: "Moscow is impressive, the active center of commerce and government. And Leningrad is beautiful, with a glorious history and beautiful art. But I really prefer Akademgorodok in Novosibirsk where the snow falls, the squirrels play, and I can do my work."

Immediately at the conclusion of this interview, a large black limousine whisked Bill off to a reception at the office of Akademik A.V. Sidorenko, the USSR minister of Geology (responsible for all Soviet Geology, including the thousands of workers in the Coal, Gas, and Petroleum industries). There were about 15 guests in his large office area for the reception, including Academician Boris Sergeev Sokolov, in whose Akademgorodok institute Bill had worked for two months during his 1975 visit, and two rather dour-faced individuals, each perched on a four-legged high-stool strategically set at two corners of the room (whom Bill assumed were KGB officers and to whom he was never introduced).

The reception was a fine event – honestly, the nicest Bill has ever experienced. As the guests were treated to hors d'oeuvres, caviar, vodka, and Armenian cognac, he chatted (with the help of a translator) with Akademik Sidorenko, during which he asked Bill whether he knew Stewart Udall, the US Secretary of the Interior and Sidorenko's American counterpart. Bill replied, "No – but his brother, Congressman Morris Udall, is now in the running for US president." (Evidently surprised, Sidorenko laughed.)

Their very pleasant conversation continued as others in the room jabbered on. Then Sidorenko, by asking a serious honest question, put Bill on the spot. His question: "You have recently spent six months in our country. What are its problems? How could it improve?" Immediately – and absolutely surprisingly to Bill – the room fell dead silent. All previous chatter ceased. With everyone in the room listening in, Bill responded by telling the Geology minister-academician what he took to be the unvarnished truth. His immediate response and the following conversation went essentially as follows:

Bill: "The Soviet Union has much of which to be proud – its history, its art, its literature, its music – and today the USSR is a strong country, but you hide your strengths from the West."

Sidorenko: "How do you mean 'hide'? What should be done?"

Bill: "Have open press conferences with the foreign reporters and show the world that the USSR has nothing to hide."

Sidorenko: "I have no control over that. I am only a Cabinet Minister, not a member of the Politburo [the Central Committee of the USSR Communist Party]. Is there anything else?"

Bill: "Well yes, and it is an even more important problem – it is your Communist Party that permits only 8% of truck drivers, doctors, geologists, or any other set of workers to be members. That's wrong, and it is not in your best interests because it means that 92% of your society has no vested interest in the success of your country. It would be far better for the USSR to have that 92% of the populace in the Party so that they would be personally committed to progress."

Sidorenko paused, thinking through what he had just heard, and then responded: "I appreciate your candor … no one around here would ever tell me such a thing."

The minister, with his aide beside him, then went to a set of drawers at the side of the room, each returning with four or five large elongate crystals balanced on their forearms. Sidorenko said, "This is a present. Take the one you like." To Bill this was like someone at a party offering him hors oeuvres and, as his mother had taught him years before, he selected the one closest to him, a four-incher. Sidorenko objected: "That's not the best one" and shoved an eight-incher into Bill's palm.

Sidorenko then, smiling, lowered the boom: "Do you know what this is?" Bill had never seen any crystal like this his life. It looked like a quartz crystal but it had eight faces, two having a bubbly texture, instead of the standard six. He hefted it; it was too heavy for glass. He then noticed two tiny wires protruding from one end and blurted out, "This is a synthetic quartz crystal." Sidorenko's response was immediate, memorable, and loud: "*Maladie'est!!*" which he exclaimed enthusiastically at full voice. All in the room applauded and actually cheered! (Bill found out later that this was what the crowd roars when a goal is scored in an ice hockey game.)

The quartz crystal Sidorenko selected was spectacular! Bill was obviously very pleased and, in his excitement, switched to Russian: "Otkuda?" (Where is this from?) Sidorenko's reply: "Zavod" (a factory). Bill then explained, "Ya ponimayu, no otkuda?" (I understand, but where?) Sidorenko's answer: "Nenazvannyy zavod" (an unnamed factory). Bill then truly understood – the location of the factory was a state secret!

Synthetic quartz crystals (which, instead of the six faces of natural crystals, not uncommonly have two extra faces not typically expressed in nature) are used in all kinds of electronic devices where precise timing is required, not only in wristwatches but in intercontinental ballistic missiles (ICBMs). What Sidorenko was telling Bill was that the Soviets were now capable of the commercial-scale production of such crystals (a process first perfected in 1950 at Bell Telephone Laboratories in the eastern United States) – a capacity that one can be certain the minister would not have revealed to him (an American visitor during the Cold War!) were he not absolutely certain that it was already well known to US intelligence agencies.

Because of Bill's ignorance about such matters, he had ventured into a highly sensitive area and, feeling a need to change the subject, he averred: "The crystal is so beautiful that I wonder whether I can get it through Customs." Sidorenko's rejoinder: "Delay kak ya … v karmane" (Do as I do … put it in your pocket!) *Ahh … Bill again lucked out!*

One final vignette from those early visits to Moscow: Bill was staying at the Ukraina Hotel, a huge imposing Stalin-era "wedding-cake" structure on the Moskva River, a few kilometers from the Kremlin city center. One sunny morning, he opted to have a stroll and see the surroundings. After a few minutes, he happened upon an elderly woman artist, Ms. Himmelfarb, who was sitting quietly alone on a park bench, painting the river scene before her. She was friendly, open, and kind and the two had a nice chat during which she pointed out where she lived – a high-rise apartment building a block away on the other side of the nearby major avenue, Kutuzovskiy Prospekt. She showed Bill her work, which he complimented – she was obviously an accomplished, experienced artist. But he also glanced at her paint palette, which was missing about a third of its colors, now only dried specks of previously used-up paint, and each of her several paint brushes was frayed to a frazil of almost nothingness. He inquired about this, suggesting that he might give her some money to replenish her supplies. With obvious dismay she turned down the offer, indicating that "Painting supplies are simply no longer available in Soviet Russia."

A scant two months later, Bill was again invited to lecture in Moscow, and he lodged again at the Ukraina. But as he prepared for that trip, he remembered Ms. Himmelfarb – with whom he had never again been in contact – and thus visited an artist-supply store near UCLA, where he purchased a large box of paints and brushes, which he carted off to Moscow. A couple of days after he had arrived, with that box in hand he walked over to the apartment building she had pointed to earlier (the address of which he did not know) and asked three or four residents, in the best Russian he could muster, to please direct him to the apartment of Ms. Himmelfarb. On the fourth try, he was successful. He knocked on the door to her apartment. She opened the door, recognized him (it had only been two months), and was terrifically surprised, absolutely astounded. He gave her his gift. She was effusively grateful. They chatted and he departed.

Two days later, Bill saw her again, sitting on the same park bench where he had first met her at the bottom of the slope that fronted the Ukraina Hotel, but now painting with her new paints and brushes. They talked and she told him that within 45 minutes after his visit, KGB officers had come to interrogate her. And then again, two hours later, several other KGBers arrived to inquire further about this contact – their mission being to find out "What did she know about this American. Why did he visit her?" In truth, she was "clean"; his visit was entirely unexpected – she didn't even know his name – and she certainly had nothing to hide. And he, too, was absolutely "clean," his motive being simply to give a gift to a decent human being that might help her in her work.

In retrospect, one can imagine that because of Ms. Himmelfarb's advanced age, the KGB officers came to visit her rather than summoning her to their downtown office as they did in the Zenaida-Leningrad episode recounted above. To the best of Bill's knowledge, this kind elderly artist was not incarcerated and had no subsequent difficulties because of his unannounced visit.

As for Bill, this fourth Russian vignette illustrates three aspects of the mid-1970s Cold War:

1. Not only the Soviet government but the Russian people as well were deeply afraid of a nuclear holocaust … as were Americans, grade-schoolers being taught to "drop and cover" with President Eisenhower having years earlier obtained the final needed vote in Congress to pass the Interstate Highway Bill by assuring the American public that it was needed for escape from cities in case of a Soviet nuclear attack.
2. Moreover, it was expected (perhaps demanded) in Soviet Russia for ordinary citizens to report any unusual seemingly "inappropriate" behavior of their friends/neighbors – pervasive nosiness that at the time was berated in the West as a reflecting a "police state" mentality.
3. During this visit, like those before, Bill was monitored (watched? trailed?) by the KGB … maybe even by the CIA. It didn't matter to him. He, like Ms. Himmelfarb, had nothing to hide.

6.5 BILL'S FRIEND, "ORIGIN-OF-LIFER" A.I. OPARIN: 1972–1976

During Bill's first visit to the Soviet Union in 1972, he met the world-renowned Russian Biochemist Aleksandr Ivanovich Oparin (1894–1980), famous for having proposed in 1924 (and in greater depth in 1936, translated into English in 1938) the first plausible scientific explanation for the origin of life (his heterotrophic "primordial soup" hypothesis). An academician in the USSR Academy of Sciences – in those days, an assemblage of nearly "god-like" figures in the Soviet hierarchy – Oparin was director of the Academy's A.N. Bach Institute of Biochemistry in Moscow, the foremost such establishment in the country.

In 1973, a year after Bill had met Oparin, the triennial meeting of the International Society for the Origin of Life was convened at Barcelona in Catalonia, Spain, organized by University of Houston Professor Juan Oró, an internationally known leader in the field and Bill's friend (who had previously tried to recruit Bill to join his Houston department). As it turned out, Oró was also a close friend of the famous Catalonian surrealist artist Salvador Dalí and had arranged for a post-conference visit to Dalí's home near Gerona, Spain, on the Mediterranean Costa Brava, so that Oparin and Dalí – both world famous – could become acquainted. Oparin and Oró invited Bill to join the group of some 10 or 12 participants.

When the group arrived at Dalí's home, Bill was first struck by the presence of dilapidated rowboat in front of his house, from the center of which emerged a flourishing Italian cypress tree (Figure 6.4). Strange! The group entered the house and Bill saw before him a large stuffed brown bear with Señor Dalí's numerous medals dangling about its neck. "Good photo," he thought. He unleashed his camera and was about to take a picture when the welcoming servant lady – an older woman attired in a black maid's uniform topped by a striking white-lace necklace – told him: "Oh, no Sir. No photographs are permitted. The Señor has his own official photographer." Bill put away his camera and the group continued on. (Years later, Bill found a photo of the current door-fronting bear at Dalí's home – after the Señor's death replaced by a festooned white-furred polar bear rather than the original brown bear; Figure 6.5.)

FIGURE 6.4 Rowboat and emergent cypress tree in front of Dalí's home.

FIGURE 6.5 Stuffed bear guarding the front door of Dalí's home.

The entourage was then escorted through a sinuous more-or-less subterranean "secret passage" – a route not taken by Oparin, who, as the guest of honor, preceded them into the home and because of his girth would have been unable to traverse this rather narrow tunnel – and were told that the passage was intended to commemorate Señor Dalí's incarceration by Francisco Franco's 1930s Spanish Fascist regime ("jail time" that Bill later learned had been only a few days). Using this tunnel, Dalí had connected three adjacent fisherman's houses (Figure 6.6), a dank symbolic underground passage presumably constructed at the urging of his wife, Gala, Dalí's legal and financial agent and muse; following Dalí's reported earlier support both of Franco and Hitler, this was evidently intended to reestablish his anti-fascist credentials with the Catalonian public. As the group emerged from that dark walkway into a small room, Bill noted a two-foot-long inflated plastic toy baby ensconced in an astronaut's space suit suspended from the ceiling above me. Strange again! This small entry room was packed with many such oddities.

The group then came out onto the interesting, if bizarre and rather small, backyard area. All of the guests were provided with canapés and glasses of pink champagne. They wandered about. Things settled down. After a while, when Bill had been talking pleasantries with Oparin and his wife Nina Petrovna, who were sitting alongside Dalí (Figure 6.7), he found himself in the truly odd position of translating their (rather mundane) conversation – Dalí knowing English, Spanish, and French but no Russian, and Oparin, only Russian. Evidently, Bill was the only other even marginal "English–Russian speaker" in the group and, thus, the appointed go-between. Despite his poor Russian, he did the best he could. After a few minutes, Oparin said:

FIGURE 6.6 Dalí's home in Gerona, Catalonia, Spain.

FIGURE 6.7 Salvador Dalí, A.I. Oparin, and Nina Petrovna Oparina.

"Sdelayte snimok nas, pozhaluysta" (take a picture of us, please). Bill looked at Dalí, who had his own official photographer (but who was absent, presumably at that time out for a smoke). Dalí nodded in agreement; Bill took a picture (Figure 6.8).

With this go-ahead from Señor Dalí, Bill then took more photos of items in Dalí's surrealistic "fairyland" backyard. As he trod about, the footing was stiff plastic grass (then a novel ground-covering, although a softer version has since been used on American football fields) and found before him an elongate reflecting pool (Figure 6.9) that ended at an Islamic cupola that housed a huge lighthouse beacon (Figure 6.10). High up on a rocky hill overlooking the garden wall was a life-sized papier-mâché statue of a camel, emblazoned on its side with the logo

FIGURE 6.8 Salvador Dalí and A.I. Oparin.

FIGURE 6.9 Reflecting pool in the backyard of Dalí's home.

"Camel cigarettes" (Figure 6.11). Strange, yet once again! He wandered about and with a bit more sleuthing discovered at the far end of a long narrow passage, too dark for photos, a bas-relief painting/sculpture of an aged nude woman whose breasts had slid down her body nearly to her knees. Immediately adjacent to this long dark tube was one of Dalí's best known creations, a pink satin loveseat (Figure 6.12) fashioned after the lips of the popular alluringly attractive movie star Mae West (Figure 6.13) – known for her memorable phrases such as "So many men, yet so little time!" – the settee surrounded by four large cardboard cutouts of Pirelli tires, two upside-down. *Wow, memorable stuff!*

In 1975, two years after this visit to Señor Dalí's home, Bill spent two months of his stint as a US–USSR National Academies exchange scientist at Oparin's institute in Moscow (during which, at the request of his fellow scientists, he gave an

FIGURE 6.10 Islamic cupola and enclosed lighthouse lamp.

FIGURE 6.11 Papier-mâché camel towering above the backyard of Dalí's home.

impromptu lecture on the similarities and differences between the US Democratic and Republican parties, in which they were much interested but about which they were highly perplexed, their sole source of relevant information being the state-controlled Soviet media). And over subsequent years, Bill has enjoyed excellent interactions with the Institute, including in 1995 having been the first to be elected as an Honorary Foreign Member of its Scientific Council – an accolade that he particularly values because it was the first decision of any kind in the Institute in which the decision-makers voted by secret ballot.

FIGURE 6.12 Dalí's famous loveseat fashioned after Mae West's lips.

FIGURE 6.13 Movie star Mae West.

After Bill had returned to UCLA following his 1975 six-month visit to the USSR, he realized that that he had a special opportunity. By then, he knew that because of Soviet government-imposed travel restrictions Oparin had little knowledge of the West and that, similarly, his Department colleagues had only limited knowledge of the USSR. "Ha," he thought, "this little problem I can solve." So, a year later, in 1976, Bill arranged through his Department and the UCLA administration to host Oparin in his lab for a two-month visit.

Because Akademik Oparin was to serve as a visiting professor, his appointment required a faculty vote. In the eyes of a few of Bill's department colleagues, Oparin's appointment was anathema – though he was not a member of the Communist Party, he was one of the three foremost scientists in the Soviet Union; he was a personal acquaintance both of Premier Joseph Stalin and of the discredited geneticist Trofim Lysenko; and his landmark "primordial soup" hypothesis fit Marxist Dialectic Materialism (providing a natural rather than a theistic explanation for how life began). Though almost all of Bill's faculty colleagues were supportive, two were adamantly opposed (one of whom opined that he "wouldn't rub shoulders with this man" and that if he ever saw Oparin in a hallway he "would turn his back and stride away in the opposite direction").

Oparin was Bill's friend (referring to Bill as his "American Grandson"), and his visit was a memorable great success for all with whom Oparin came in contact. There were many UCLA faculty members from diverse disciplines (not only the sciences but from the humanities and social sciences as well) who wanted to meet him, so many that Bill divided them into manageable groups, and every Tuesday evening, his wife Julie and he hosted a cocktail party at their home (which, interestingly, Bill's departmental colleague nay-saying "walk-the-other-wayer" ultimately asked to attend).

6.6 OPARIN TEACHES UCLA STUDENTS: 1976

Once a week during Oparin's visit, Bill staged a luncheon in his lab to which he invited 10–12 UCLA undergraduates and graduate students to meet and interact with this internationally acclaimed star (Figure 6.14). On one of those occasions, Bill asked Oparin to recount (translated from Russian to English by Bill's Berlitz teacher) how he had become interested in the problem of the origin of life – a story that Oparin had previously told Bill that would inspire the students, beginning when Aleksandr Ivanovich was a high-school student of about the students' same age. The following is Oparin's story.

Dating back to 1900 and before, virtually all scientists who addressed the matter had assumed that Earth's earliest forms of life must have been plant-like microbes, simply because animal-like organisms could not have survived unless there was plant-stuff already available to eat. Oparin, however, had a different idea. Beginning in 1910, when he was a high-school student in Uglich (a small town on the Volga River northeast of Moscow), Oparin carried out simple growth experiments on plants, spurred largely by the writings of Moscow State University (MSU) Botany professor K.A. Timiryazev. From these early experiments and Timiryazev's textbooks, Oparin came to realize that plants build sugar by photosynthesis, which they then break down to provide energy to fuel their cells. Thus, he reasoned, because plants both make *and* consume organics – whereas animals can only consume them, not build them from scratch – the way that plants live is actually more complicated than that of animals. In terms of energy production, plants are therefore biochemically more complex, animals more simple.

In Oparin's last year in high school, 1912, he like his classmates took an all-Russia national examination (not unlike the current Scholastic Aptitude Test, the SAT, used today by universities in Australia, Canada, the United Kingdom, and the

FIGURE 6.14 A.I. Oparin and UCLA students.

United States) in the hope of being admitted to Moscow State University (MSU), the foremost university in the country. He and one other from his graduating class were admitted, and their science teacher arranged to take them on a train from Uglich to Moscow to have a first-hand look at MSU (just as many students do before enrolling at American universities). The group missed the first train and arrived in Moscow in the mid-afternoon, giving Oparin the opportunity to attend only one university lecture. For that he selected a late afternoon presentation by the aforementioned Professor Klimet Arkadievich Timiryazev (1843–1920; Figure 6.15) – known by Oparin as the author of two botanical textbooks that Oparin had read – a nationally renowned botanist of acclaimed oratory skills and the outspoken leading proponent of Darwinian evolution in Russia.

As serendipity would have it, this was near the end of Timiryazev's academic career, and he devoted his lecture to recounting how he, Timiryazev, had become a confirmed Darwinian evolutionist – an exceedingly unpopular view in a country dominated by the teachings of the Russian Orthodox Church (as it was then also in the United States, well illustrated by the nearly contemporaneous 1925 "Scopes Monkey Trial"). Timiryazev's father had also been a professor at MSU (an understandable familial descent then typical throughout academia) who, upon Timiryazev's receipt of a doctorate degree in 1870, had funded his son's visit to England to visit Darwin (the *Origin of Species* having been published 13 years earlier). Although the 27-year-old Timiryazev had received no reply to the letter he had written to Darwin requesting an audience, he journeyed off to England.

FIGURE 6.15 Soviet botanist Klimet Arkadievich Timiryazev.

The young Timiryazev arrived at Darwin's home, "Down House" in the rolling hills ("Downs") of Kent, southeast of London; rented a room at the Queen's Head Pub down the hill on Luxford Road; and day after day, for a week, sat patiently on the stoop at the entrance to Darwin's house waiting to see the great man (who, it should be recalled, now at age 61 was beset with debilitating maladies). The encounter finally came, and the two walked the long gravel-paved tree-lined "sand-walk," Darwin's so-called thinking path at the rear of the property, as Darwin taught about evolution. Timiryazev became convinced – Darwin's evolution was reality.

Young Oparin, an 18-year-old about to graduate from high school, was enthralled by this first encounter with MSU and, particularly, by Timiryazev's lecture. His take-away lesson from Timiryazev's lecture, as he told the students around the lunch table, was that "Darwin had written the book on the evolution of animals and Timiryazev had explained the evolution of plants, but the first chapter in life's history – how life began – remained unknown."

For the next few years, as an undergraduate at MSU, Oparin pondered the problem and came up with the notion that because life must have evolved from simple to complex, as Darwin had suggested – and because plants were more complicated than animals in their ability to both make and then use organics (Timiryazev's teachings backed by Oparin's high-school experiments) – simple single-celled "animal-like" (i.e., heterotrophic) microbes must have evolved before the development of more complicated photosynthesizing forms of life. Moreover, because plants are the source of Earth's oxygen, a gas that would therefore not have been present before they came into being, life's origin must have occurred in an oxygen-free setting. So far, so good – it all seemed to fit together. But if "plant-like" photosynthesizers had not yet evolved, what would the earliest "animal-like" microbes eat and from what had *they* evolved? Oparin's solution: a "primordial soup" of organic compounds formed in an oxygen-free environment by reactions powered by sunlight (i.e., UV radiation from the early Sun) interacting with gases in Earth's primitive atmosphere, a non-biologically formed organic-rich broth that would have provided the chemical building blocks for life's emergence as well as the food required for its survival.

In 1918, a year after Oparin's graduation from MSU, he wrote up his views on life's origin and submitted a manuscript. His paper was summarily rejected. At the time, how life began was almost universally regarded to be a theological rather than a scientific question. And even though the overthrow of the Romanov Dynasty and the abdication of its final Emperor (Nicholas II) during the October 1917 Bolshevik Revolution had occurred a year earlier, Oparin's manuscript fell under the purview of the Tsar's censors, who were still in place, and their adherence to the doctrines of the Russian Orthodox Church continued to hold sway.

Over the next few years, Oparin refined and clarified his ideas, which were first published in a 1924 (in Russian) in a rather slim volume that he referred to affectionately as his "little pamphlet." He told the class that the six-year delay had been of great help to him because it had enabled him to fortify the bases of his argument, most notably by permitting him to take into account the then-new discovery of hydrogen and methane in the atmospheres of Jupiter and the other giant planets, which led Oparin to suggest that Earth's early atmosphere would similarly have been hydrogen-rich – composed, in his view, of methane, ammonia, hydrogen, and water vapor – the source of the raw materials needed for the formation of a life-generating and -supporting primordial soup. Twelve years later, in 1936, he published an expanded discussion of his hypothesis (again in Russian), which, with its translation into English in 1938, set the stage for modern scientific understanding of the origin of life.

There are a few other matters not mentioned by Aleksandr Ivanovich around that lunch table that deserve comment.

1. His 1924 solution to the central problem of the origin of life – his "primordial soup hypothesis" – had no experimental backing and flew in the face of then-accepted dogma. Non-biological ways by which organic compounds can be produced were entirely unknown, and the chemistry of organic matter ("organic chemistry") and the chemistry of life ("biochemistry") were regarded as essentially synonymous. Indeed, it was not until 1953, three decades later, was it shown by Stanley L. Miller (Figure 6.16) and his professor, Harold Urey, that the amino acid building blocks of proteins can be synthesized by wholly non-biological mechanisms in just the sort of strongly reducing environment that Oparin had envisioned. Within a few years after this breakthrough, as others joined the fray, a host of other organic compounds were shown to be synthesized by the same prebiotic mechanism. Oparin's basic notion had been confirmed, as it has now been repeatedly over many decades.
2. In 1929, only five years after the publication of Oparin's idea, the English biologist–geneticist John Burdon Sanderson Haldane (1892–1964; Figure 6.17) – like Oparin, a highly respected scholar – published a short

FIGURE 6.16 American origin-of-life chemist Stanley Lloyd Miller.

FIGURE 6.17 English biologist–geneticist John Burdon Sanderson Haldane.

paper that presented essentially the same ideas that Oparin had previously suggested. Though Haldane had independently arrived at these notions, he later acknowledged Oparin's priority of the concepts. In that era, such unfamiliarity with the international scientific literature was typical – at least partly due to "language problems," those outside the reaches of the USSR having little knowledge of the Russian language. History shows that such lack of knowledge carried over well into the mid twentieth century, illustrated, for example, by Stanley Miller's comment to Bill that, when he set out to do his pioneering work on the synthesis of organic compounds under simulated early Earth conditions, neither he nor Harold Urey had even the slightest inkling of Oparin's seminal hypothesis (it having been called to their attention only by a reviewer of their 1953 manuscript). Thankfully for the scientific community, this deficiency has been rectified and the community interconnected by the universal use of English for major scientific publications … in Russia, China, South America, the countries of continental Europe and elsewhere in which English is not a native language … a result, one can easily imagine, of the 1815–1914 British "imperial century," when English was spread worldwide.

3. As Oparin privately told Bill, it was his view that well-constructed theory is of primary importance – that "a good theory can stand alone" – and that supporting facts (for example, the decades-later Miller-Urey experiments), "nice as they can be," are of secondary import. Although a similar view of the primacy of theory may be shared by some academic disciplines, in most fields of science, well-established facts reign supreme and theoretical concepts are constructed secondarily, their purpose being to link the factual observations into a coherent whole.
4. Like it or not, cultural differences and international politics play a role in the acceptance of ideas in science (even today). The 1917 rise to power of Communism in Russia was viewed by the Western world as a fearful threat. Though Oparin was never a member of the Communist Party, as a star Soviet scientist he was personally acquainted with USSR premier Joseph Stalin. Haldane, in turn, had become an avowed Socialist during World War I (1914–1918), became an open supporter of the Communist Party in 1937 (writing numerous articles for the *Daily Worker*), and joined the Communist Party of Great Britain in 1942. Moreover, the naturalistic rather than theological explication of the origin of life shared by Oparin and Haldane and their common idea of an evolutionary continuum from inanimate to animate, from simple to more complex, fit Communist Marxist Dialectic Materialism. Because the views of Oparin and Haldane both were thus widely regarded to be "politically suspect," their Marxism-consistent concepts were, in many quarters of the Western scientific community, assumed not to be trusted.

What was the value of Oparin's story to the students at that lunch table? Of course they heard it first-hand from an internationally acclaimed "Great." For them, no doubt, this experience has been permanently memorable. But for Bill, it stands out primarily because it illustrated the straight-line descent from an "old" Darwin (then aged 61) to a young Timiryazev (then 27); from an aging Timiryazev (69 when Oparin heard him lecture) to the young Oparin (then age 18); and, lastly, from an elderly Oparin (then aged 82) to all around that lunch table. Many in science acknowledge that we today "stand on the shoulders of giants" (an oft-quoted aphorism attributed to the

1675 writings of Isaac Newton), and to Bill, this four-step human connection from Darwin to Timiryazev to Oparin to us proves the point. Think about it – Darwin, whose contributions most students regard as dating from really, really long ago – is by this genealogy their intellectual great-grandfather, Timiryazev their grandpa, and Oparin their father, a much shorter link of human interactions than many would have imagined. Moreover, as Bill much hoped, several of the students listening to this tale were inspired by the youthful Oparin's insight, imagination, and perseverance – if he could be such a game-changer, perhaps they also could aspire to scale such heights – an inspiration that some still regard to be among "the most memorable experiences" of their days at UCLA.

6.7 A.T. WATERMAN AWARD: 1977

That same year, 1976, was the bicentennial of the founding of the United States of America. To commemorate the event, the US National Science Board established the Alan T. Waterman Award (Figure 6.18) to be presented annually to one "Outstanding Young Scientist or Engineer in the United States under age 35." In 1977, Bill was selected as the second recipient. He had previously never heard of the award, but it has since risen to its intended prominence, the United States' highest honorary award for young scientists. He learned later from a member of the Board that he had been ranked highly both by their Physical Sciences and Life Sciences panels and that the Board regarded this as their first (and perhaps last) chance to select a practitioner of paleobiology, a field that, because of its inherently interdisciplinary character, is populated by late-bloomers. Now, four decades later, he has continued to remain the award's sole Paleobiology or Geoscience recipient.

The award carried with it the munificent prize of $150,000 (now one million dollars), no strings attached. But, as Bill was told, his receipt of the award presented a problem to UCLA chancellor Charles E. Young. The Science Board insisted that

FIGURE 6.18 US National Science Board's Alan T. Waterman Medal.

because this was a prize, not a grant, it should not entail standard university overhead fees (then ~35%, now 54%). The Chancellor balked, concerned that this would set an unwelcome precedent. However, after considering the matter further, he acquiesced, evidently figuring that this was a "one-off" that would never happen again. If so, he was mistaken. In 2008, UCLA's exceedingly accomplished young mathematician Terrance Tao was the award's 32nd recipient. (When Bill was informed of this, he wrote a short note to Professor Tao informing him of the established precedent in the hope that he, too, could avoid the overhead charges).

6.8 A VISIT TO CHINA – THE SECOND MAJOR LIFE-CHANGER: 1978

Like Bill's 1961 introduction as a second-year college student to Darwin's Dilemma, the "missing" pre-Cambrian record of life, his 1978 visit to China was a major life-changer. Thanks to contacts established by Yale Botany professor Arthur W. Galston, the Chinese government invited a 10-person delegation representing the Botanical Society of America (BSA) to spend a month touring and lecturing in the People's Republic of China (PRC). Bill was selected to be the delegation's paleobotanist. This was the first such invitation ever extended to an American delegation, and though it occurred after an earlier short episode of "ping-pong diplomacy," it preceded normalization of United States–China diplomatic relations that later ended three decades of official estrangement. (In retrospect, it seems obvious that the delegation's visit represented a first safe step in this process, for which Botany qualified as it presented no obvious military or high-tech threat to the PRC).

For Bill, this trip was a superb introduction to Chinese culture, history, and science, which in 1981 and again in 1982 he followed up with research visits during which he discovered the oldest fossils then known in all of Asia and helped China to become a world-class contributor to Precambrian Paleobiology. But it was this initial 1978 trip to China that was a true life-changer because it was during this trip that he met fellow delegation member plant biologist Jane Shen-Miller, whom he wed two years later and who has been his long-time compatriot, supporter, and the love of his life ever since – and, he is pleased to note, a highly honored world-class scientist (2018 recipient of the Life Time Achievement Award of the American Society of Plant Biologists).

Unfortunately for Bill, however, his 1978 trip to China did not sit well with his aunts in Iowa and New Mexico. In their eyes, his previous sojourns to the USSR pegged him as a possible "Pinko," but this visit to the PRC, the other "Great Communist Power," was just too much to stomach – they both stopped talking to him.

Both of the follow-up 1981 and 1982 multi-month visits to China were facilitated by the selection of both Bill and his wife, plant scientist Jane, as Chinese Academy of Science (CAS) Visiting Research Scientists. For the first two months of the 1981 visit, they carried out our research at the Botanical Institute of the Academy, then housed on the grounds of the Beijing Zoo, a couple of miles distant from their lodging at the Friendship House Hotel Complex. This massive park-like complex, a walled-in cluster of 10 or 12 large hotel buildings, was originally built to accommodate (and isolate) Soviet advisors who in the 1950s aided in the rebuilding of China after the Chinese Communist Revolution of 1945–49, perhaps most importantly providing needed expertise for the production of carbon-infused steel.

During both of these visits, the Schopfs were housed in the main building of the complex (Figure 6.19), and when they first arrived, Bill was astounded to see that the reception desk and the entire hotel lobby were faced bottom-to-top (flooring, pillars, reception desk, etc.) by deep red (iron-rich) specimens of Precambrian stromatolitic limestone (Figure 6.20). He had seen such stromatolitic flooring earlier, in 1978, when their delegation arrived in a hotel in Canton (now Guangzhou), and Bill kneeled down on the lobby floor to take a picture using a crisp 5¥ (*Yuan*) note for scale – and a local Chinese gentleman loudly chastised him for "dirtying the People's money" – but the Friendship House lobby was the most splendiferous display of stromatolites he had ever seen.

Because the Schopfs were official guests of the CAS, the Botanical Institute sent a hired car and driver each morning to ferry them to work. After a week or so of this kind service, Bill asked the driver how much it was costing the Institute. Astoundingly, the weekly sum he mentioned for a mere three days of his service was equivalent to the average monthly wage of a Beijing citizen! To Bill this seemed wrong, not because of the kindness of the Academy's effort to ease their transport to the institute, but because he regarded it as unnecessary and far too costly.

That morning, after hearing this from the driver, Bill inquired of his new friends at the Institute as to whether anyone had an unused bicycle that he could borrow.

FIGURE 6.19 The author's wife, Jane, and son, Jamie, in front of the Friendship Hotel, Beijing.

FIGURE 6.20 Stromatolitic floor of the Friendship Hotel lobby, Beijing.

One did and brought it in the next day. From then on, Bill rode that bicycle to work – which, for him, was lots of fun. Think about it. The trip was not very long and, as a youngster, he had ridden his bike five or six miles a day, not just a measly two. Moreover, at that time, everyone in Beijing rode bikes to work, transport by automobile being limited to Chinese dignitaries and foreign guests, so the roads were clogged by bicycles, not cars. On the road to work there were several stoplights, where hundreds of bike-riders would bunch up, waiting for the light to change. They, in their blue or gray "Mao uniforms," would glance quizzically at Bill – a white-faced foreigner wearing a coat and tie – and wonder in great surprise. OK with him – they were interesting to him, too, and he was in there pedaling with them because it seemed the right thing and, besides, it was excellent good fun. For three weeks, he managed to get away with this ploy when he was then called in to the institute director's office to explain his actions, the director's concern being that Bill might be hurt. They talked; Bill explained; then the director agreed: "Well, you haven't been hurt yet and you are saving us an enormous amount of money for which I thank you."

Bill discovered during this 1981 visit that, to his surprise and great pleasure, Precambrian stromatolites – strikingly handsome if vertically sliced to expose their characteristic fossil microbe-formed stacked layers – were used widely in Chinese architecture. One Saturday afternoon, accompanied by two geologists from the Academy's Nanjing Institute of Geology and Paleontology (NIGPAS), whom they had met in 1978, Bill and Jane ventured down to Tiananmen Square in central Beijing, first visiting the National Museum of China on the square's eastern edge. As they departed the museum, Bill noted a huge building far across the square – some five football-fields distant – "the Great Hall of the People," which was fronted by a phalanx of imposing pillars that from that distance he guessed, but had no way to know, might be stromatolitic (Figure 6.21).

The four of them trundled across the square and Bill walked up the guard-posted visitors' ramp. He had a look at the columns nearby and confirmed that they were indeed made of stromatolitic rock, but saw that they were dirty and smudged, not

FIGURE 6.21 Great Hall of the People, Tiananmen Square, Beijing.

worth a photo. So he then walked down the lengthy front of the building and near the far end found a clean column to photograph to show the students in his classes (Figure 6.22). Like all of the columns, its base had been constructed from the deep red stromatolite-rich limestone of the 850-million-year-old Northeast Red Formation (from near Dalian, Liaoning Province) and the facing on the upper portions of the columns from 1,450-million-year-old gray stromatolitic limestone of the Gaoyuzhang Formation (quarried near Jixian, about 80 km east-northeast of Beijing). Within about 90 seconds, three Chinese uniformed guards surrounded him and ordered him to leave this "unauthorized area." He of course immediately complied (and, thankfully, they did not confiscate his film). After he had retreated down the ramp, he noticed that the two NIGPAS escorts were nowhere to be seen – they had escaped the scene (and did not rejoin Bill and Jane for the following two days), wanting nothing whatever to do with Bill's encounter with the police.

There are a couple of other Precambrian stromatolite–related stories from this 1981 visit. One morning, as Bill was riding his borrowed bicycle to work, he glanced off to the right and saw a "Peoples' Park," a dirt-floored children's play area without a scrap of grass in which there was a playground slide that caught his eye. It, too, was composed of slabs of stromatolitic (Gaoyuzhang) limestone encased by metal siding … presumably a cost-effective technique for its construction (Figure 6.23). And one day at the institute around lunchtime, after his lab colleagues had departed to their homes for their noon-day meal and an early afternoon rest (as was the norm in those days), he received a phone call from Jane, who was doing her own work in a different institute lab, telling him that her "lab benches are made of stromatolites." Bill was skeptical – what did a plant biologist know about stromatolites? But to placate her (and keep peace in his family) he walked across the campus to have a look.

FIGURE 6.22 Stromatolitic pillar, Great Hall of the People.

FIGURE 6.23 Stromatolitic playground slide at a Beijing "Peoples' Park."

She was, in fact, absolutely correct – as he learned before, then, and since, she is very perceptive!

A native Chinese, born in Shanghai (and later having been educated at Washington State and Michigan State Universities), Jane is fluent in Mandarin. So, Bill suggested that she call around and find out where her lab benches had been manufactured. She did, and a few days later, they were invited to visit the Beijing Peoples' Marble Factory (whose workers, as Bill recalls, composed a part of "Factory Production Brigade #10"). The director and his deputy greeted them warmly. As the group sipped tea, Bill gave them a short chalk-and-blackboard non-technical lesson about how stromatolites are made (structures they referred to as "wreath-ring rocks," in Chinese, *huaquan shí*) – a subject about which they had no evident interest.

The deputy director then took Bill and Jane down to see the factory floor – a huge, truly impressive area (larger than UCLA's 18,000-seat basketball pavilion) in which a dozen water- and sand-fed abrasive saws were cutting through huge quarried blocks, three to four meters on each side, of Precambrian stromatolitic rocks. And off to one side were some 50 carefully arranged horizontal stacks of sliced stromatolitic limestones, each stack organized such that the cut faces of the slabs showed sequential series either of cross-sectional (horizontal) or vertical views of the stromatolitic structures – precisely the information needed for a paleobiologist to prepare accurate reconstructions of their three-dimensional morphology. In short, the sectioned specimens represented a treasure trove that was ripe for detailed scientific study (an observation that Bill later passed along to several of his Chinese colleagues who, unfortunately, never followed up).

The deputy director then took them into a yard behind the factory, where broken or otherwise unusable slabs had been discarded. There Bill discovered a large, one-by-two-meter beautifully deep red slab packed full of the stromatolite *Baicalia* (quarried from the 850-million-year-old Northeast Red Formation) from which a sizable corner had broken-off. Seeing his interest, the deputy explained that this slab, along with five others, had been prepared in 1976 to be used as facing to enclose the entrance of Chairman Mao's Mausoleum, at that time under construction, but a few days before their scheduled installation, it was realized that their limestone (rather than crystalline marble) composition would rapidly degrade in Beijing's pervasive acid rain. So, despite their deep "Communist Red" color, they were not put in place, the five unbroken slabs having been sold some years earlier, with this last slab having been discarded because of its large missing corner. When Bill then inquired how he might purchase it and have it delivered to the United States, he discovered that there were three different price categories, depending on the nationality of the buyer: full-price for a foreigner (in this case, him); half-price for an "Overseas Chinese"; and quarter-price for a Chinese citizen. Because Jane qualified for the 50% discount, she officially purchased the slab, which Bill arranged to be cut down, squared-up, re-polished, and shipped to them in Los Angeles, where it is embedded in the entryway to their home.

In the final few weeks of this 1981 visit, rejoined by their NIGPAS colleagues, they drove eastward from Beijing to the farm country near Jixian to collect rocks from the officially designated (and radiometrically well-dated) "Type Section" of Chinese Proterozoic (550–2,500-million-year-old) Precambrian strata. During this first visit to the Jixian farming commune, Bill had several memorable experiences, among which one particularly stands out. The four in the group, the two NIGPAS geologist–paleontologists and Bill and Jane, were collecting rock samples from an exposure on the side of a dirt farm road. As their work progressed, Bill glanced across the road to a nearby farm field and noticed an elderly lady hanging clothes out to dry on her side-yard clothesline. After some four or five minutes, she returned, took down the clothes, and then went back into her house. Oddly, it seemed to him, she repeated this activity over and over for the following 20 or 30 minutes. It was clear to him that she was simply curious about the visitors – who were they, what were they doing, why were they there?

When they finished their work, they walked over to meet her, and Bill introduced himself and the members of the group. She smiled, she was welcoming, and she invited them into her home, a one-room cement-block structure (with a poured concrete bed, in Chinese a "*kang*"), only a year or so old; like others throughout the commune, this had replaced the previous dilapidated wooden houses, and she was greatly pleased with it. She was also quite proud of her son, showing the visitors his picture in a smart-looking Chinese army uniform and telling them (with Jane translating for Bill) that for the first time in her son's life he now had "decent clothes." They all then had a pleasant conversation, during which she told these strangers that she, like all in the commune, was "happy but not wealthy." Her attention and remarks were primarily directed toward Bill – chiefly, he imagines, because he was no doubt the first "white-faced foreigner" she had encountered since the 1940s.

She went on to tell her guests about her life and that she now had only one hen, a scrawny, aged chicken that laid one egg each week. Then – in an extraordinary act of kindness and generosity – she asked Bill if she could cook the egg for the group. He, of course, was deeply touched, but he also knew that would have been inappropriate (indeed, in his view immoral) for them to have accepted her offer and thus "steal" her single weekly egg. He declined as graciously as he could manage. She then offered to prepare "White Tea." Bill thought to himself, "it will be boiled, so we are not likely to become ill," and her generosity, kindness, and openness simply *had* to be rewarded. Having not the slightest idea of what she meant by "White Tea," he told her, "Oh yes – that would be wonderful!" ("White Tea" turned out to be boiled water – no tea leaves, no flavoring, only boiled water.)

To Bill, this generous, kind, open, and honest woman reinforced what he takes to be a Great Truth: All humans, globally – and despite their differences in education, opportunities, family background, and community/national culture – are fundamentally alike. We all are social primates, and we all have the same needs and cares as members of any of the other disparate human "tribes." (A few years ago, Bill recounted this story and its moral tag to his freshman-sophomore General Education class, after which two young women, both of Chinese descent, came up and thanked him – with tears in their eyes.)

Before the group had headed off to do this field work, Bill had asked his NIGPAS colleagues to obtain geological maps of the Jixian area, with which he was not acquainted. They tried but failed because such maps were officially regarded as a "State Secret" (evidently because of the nearby location of an army base, of which neither they nor he had prior knowledge). After the group's nice get-together with the kind aged commune farm-worker, they headed down the farm roads and at one locality collected rocks that to Bill seemed particularly promising for the lab studies he planned to do. He, of course, wondered where else these strata might also occur and what roads might lead to such outcrops, information that he would normally have obtained from (the non-available) geological maps. The group then journeyed into a nearby side-valley, and while the three others in the group collected rocks on the valley floor, he climbed a hill high above and using binoculars made a sketch map of the strata and roads of the local area.

As Bill ventured back to the group and got closer and closer to the valley floor, he heard a high-pitched yelling: "Jiaoshou, Jiaoshou [Professor, Professor] come down, come down." After he finally did, cheered on by these repeated unexpected outbursts, he looked around to find their source. Nothing – no one was there. He then saw a small building, which he imagined to be the local grade-school, and adjacent to it an area devoid of grass (much as he had previously seen at the children's play area in Beijing), from which emerged five well-separated trees. Bill sat down in the playground at the middle of the little grove of trees and waited. From time to time he glanced around, and after a few minutes, he saw a small head, then another, peeking out from behind one of the surrounding trees. He then waved in the direction of the trees, gesturing for the youngsters to come sit with him and yelled out "come out, come out, wherever you are." They did. And then their teacher emerged from the schoolhouse and introduced herself, and they all went into the one-room commune school. Once all were in place, Bill said a few words during which, he hoped, the Chinese youngsters discovered that white-faced foreigners are, at least not always, the then much-feared "running dogs of colonialist/capitalist empires."

At the conclusion of this 1981 China visit to Jixian, the four-person field team was invited to a fine farm dinner hosted by the commune elders – which they much enjoyed, as did their visitors – the elders primarily because the special food (including the traditional meal-ending fish course, not normally available at the commune) was provided and paid for by the Academy of Science, the sponsor of the Bill and Jane Schopf group visit.

Bill and Jane returned a year later (now, one of their dozen or more trips to China). On this 1982 visit, they continued their field work and Bill brought along for the local grade-school a meter-diameter thick-skinned inflatable playground ball and a *National Geographic Atlas of the World* – so that the children would have some way to find out where this white-faced foreigner (*laowai*, "old outsider"; in Chinese slang *dabizi*, "big nose") had come from. At the dinner that they knew would conclude their visit, Bill presented the commune elders with a meter-square Landsat image of their entire commune that showed it extending at its northern edge to the Great Wall of China. They much appreciated this gift (as Bill silently thought to himself that though the local roads and geology of the region might be regarded by the Chinese authorities to be a "State Secret," they would certainly not be so to anyone who could afford the $45 cost of a Landsat image.)

Although the 1981 and 1982 to China were memorable to Bill – and in retrospect turned out to be pivotal to the development of Precambrian Paleobiology in Southeast Asia – it was the initial 1978 trip that changed his life, principally, as previously mentioned, because it was during that trip that he met fellow-delegation member and soon-to-be wife, plant biologist Jane Shen-Miller. When they first met, they both were married – she for 17 years, him for 13 – so the path to their subsequent union was not without its painful moments.

6.9 DIVORCE AND REMARRIAGE: 1978–1980

Bill was not unhappy in his first marriage – he was devoted to his son and his former wife. Their home-life was pleasant, placid, without strife, and he and son Jamie were wonderfully great pals. In the year or so after Jamie was born Bill got up in the middle of the night and changed his diapers; regularly prepared his "bottle and formula"; and they played and played and played together – Bill singing and slinging Jamie's little arms from side to side, to-and-fro (a gentle game that Bill thought might help develop his coordination). Later, when Jamie was of school age Bill prepared his daily lunch, walked him down to meet the school bus and coached his "mini-league" soccer team. Wife Julie, however, had a tough time adjusting to the arrival of baby Jamie which they both had been greatly looking forward to. Her mother came out from Massachusetts to help the new parents get through the first two or three weeks. However, when her mother departed Julie felt abandoned and it fell to Bill to meet their new needs (diapers, bottle, formula, and so forth). By all accounts, the birth itself was not the problem. Rather, it was the aftermath – which can evidently be chalked up to Postpartum Syndrome, a common reaction to childbirth (exhibited by an estimated 15% of new mothers) resulting from hormonal changes and not infrequently accompanied by one, the other or both parents' initial lack of confidence in being able to cope with the new-found responsibility.

While Bill was off at work Julie spent her days tending the small garden at their home and having coffee with the wives of his colleagues who lived in the neighborhood. Over the following few years all but one of the wives found jobs, mostly at UCLA, and Julie was once again left (almost) alone – not least because of Bill's seven-day work-week. The up-side of his work ethic was that he rose rapidly in the professoriate, but the down-side was that it took a toll on his marriage. Ultimately Julie turned to Kriya Yoga meditation, an ancient technique of energy and breath control (pranayama) in the Hindu tradition during which the practitioner meditates and softly chants for hours at a time in an effort to find "inner peace."

After 13 years of marriage, the bloom was off the rose. So, following Bill's 1978 China trip he and Julie divorced and she and Jamie departed to be near her relatives in Washington State. The separation was about as "amicable" as such things can ever be – neither of them ever uttering anything unpleasant about the other – but it was the most difficult, heart-breaking episode of Bill's life. Julie later joined a religious commune in northern California and is happily remarried.

Thankfully, son Jamie has succeeded admirably. After high school, he was admitted to Oberlin College (of which his mom and dad were both graduates). Jamie did well at Oberlin, graduating as an Honors Student in Political Science. He then went on to pursue a doctorate in Political Science and International Affairs at the University of California, San Diego, studying under the tutelage of Professor Stephan Haggard (a world-class leader in studies of international politics, especially Korea–Pacific studies). The research resulting in Jamie's 2004 doctoral thesis was difficult and courageous, requiring the release from the South Korean government of previously sealed corruption- and bribe-related National Assembly documents, which he succeeded in obtaining with the help of lawyers from the Korean equivalent of the

American Civil Liberties Union, supported by the then newly amended Korean Freedom of Information Act. He, his Korean wife, and their two children now reside in southeast South Korea, where he is a professor of International Affairs, Politics and Diplomacy at Keimyung University in Daegu. He has had a good run in academia. *Bill cheers his success!* (And, besides, thanks to Jamie's family, Bill is now a two-time grandpa!)

The story of how Bill met and became enamored with fellow BSA delegate Jane Shen-Miller – to whom he has now been married for 38 years – during their short 31-day 1978 trip to China is simple and to him rational (though some have dubbed it "romantic").

During the trip, each of the members of their visiting group had to give a lecture. Bill's turn came when they arrived in Nanjing (where members of the Nanjing Institute of Geology and Paleontology still remember him as their first "foreign visitor"). His talk had been well prepared, but it dealt with Geology, Paleontology, Biology, Biochemistry, and Organic and Isotopic Geochemistry. Their Chinese hosts were unable to locate a translator who could understand the English words in the diverse fields, so Jane (then a program officer at the National Science Foundation in Washington, D.C.) was assigned the task. By then, from his earlier translated lectures in the USSR, he knew the ropes – short, simple sentences that would then be converted into Chinese. About a third of the way through the lecture he was talking about carbon isotopic fractionation during the process of photosynthesis, the fossil record of which extended to more than three billion years. Fellow delegate Jane began to translate his short sentence and then – in front of an audience of some 500 – she turned to him on the lecture stage and blurted out "Is that really true?" He replied, "Of course. Please translate." She persisted: "How do you know?" He smiled and said: "You'll soon find out – please translate." She did. Bill decided, then and there, that "this person has a mind. She thinks. She wonders. She questions. She is first-rate!"

Jane's divorce was no easier for her than Bill's was for him. But for him, at least, becoming married to this sterling plant biologist was the best thing he ever did. In truth, among the plant biology community, she is appreciably more "famous" than is he, in part – but by no means entirely – based on her unprecedented discovery in 1995 that the seeds (in botanical nomenclature, "fruits") of the Sacred Lotus (*Nelumbo nucifera*) can remain viable for as long as 1,300 years, the world's oldest directly dated living seeds!

Jane's discovery of this world record–shattering seed viability – a good example of the insight and perseverance required of excellence in science – is of considerable interest. On one of their later trips to China, she obtained from the Beijing Botanical Institute seven Sacred Lotus fruits that had been exhumed from a peat bed at the village of Xipaozi near the town of Pulandian, northwest of Dalian (in a region once referred to as Manchuria). The area of Xipaozi at that time was covered by farm fields that, millennia before, had been a large lake filled with Sacred Lotus planted by Buddhist monks as Buddhism spread northeastward from India across China and south into Korea. As a result of earthquakes in the region, the lake drained into the nearby sea, and over subsequent centuries, the peat bed became buried by a two-to-three-meter-thick layer of windblown loess derived from the Gobi Desert and northern China. From time to time, dating back to at least the early 1900s, shiny relatively large ("Super Colossal" olive-sized) individual lotus fruits came to be exposed at the top of the tilled fields, where they were collected by the local farmers and prized by the farmers' wives, who stored them away in small cloth bags.

During PRC Chairman Mao's (disastrous) "Great Leap Forward" of 1958–1961, exhumed Xipaozi fruits became appreciably more abundant. A principle goal of Mao's plan was to make China industrially and agriculturally self-sufficient (within a scant five-year period), and farm communities, some 85% of the Chinese populace then being rural, were encouraged to be self-reliant. To accomplish this feat, the farmers were urged to construct backyard blast-furnaces with which to make the iron and steel tools needed for their work. To fire the furnaces, the Xipaozi farmers needed fuel, and the peat bed underlying their crop-land could solve the problem. So the farmers laboriously overturned the beds, exposing the peat and its treasure of lotus fruits.

As the "Great Leap Forward" miserably failed (soon replaced by the similarly disastrous widespread starvation of the "Cultural Revolution"), the furnaces at Xipaozi disappeared. In the ensuing three decades, from the 1960s to the 1990s, lotus fruits continued to weather-out from the overturned beds, which the farmers gathered and some simply ate (virtually all parts of the lotus plant being edible). But for others, the fruits remained valued "collectibles," prized because of the cultural and religious significance of lotus, its beautiful large flower (Figure 6.24) serving as the "throne" of the iconic "Sitting Buddha" statues.

Jane and Bill have returned to Xipaozi several times since those early trips – in 1996 accompanied by a French television crew that filmed her for a one-hour "*Tele-Image* Special" featuring her work (that, as it turned out by sheer chance, the two of them watched in their hotel room during a visit to Paris the following spring). Over the subsequent two decades, the Xipaozi landscape changed markedly. The farm fields of the 1990s and before were paved over as a new superhighway cut through their midst, and the Chinese "building boom" brought new housing and industry and

FIGURE 6.24 Flower of the Sacred Lotus, *Nelumbo nucifera*.

a great upsurge in the local population. Today, farming has all but vanished, and the once fairly abundant lotus fruits are no longer collectable.

In Jane's first experiments, she germinated the fruits under green light (to block out photosynthesis and the incorporation of carbon into the emerging seed leaves) and then Carbon-14 dated the germinated fruits in Nobel Laureate Willard Libby's laboratory in the UCLA Geology Building (Libby having been the founder of the ^{14}C radiometric dating technique). For the dating part of the project, this worked well, but at that time the technique required a relatively large amount of carbon for accurate age-measurement and, thus, the incineration of an entire fruit. And this, in turn, made it impossible to grow a dated plant from seedling to adult and thus study its growth characteristics and genetic composition. In subsequent years, Jane overcame this problem using a new dating technique (accelerated mass spectrometry) that requires only a minuscule amount of carbon from the outer coat of a fruit. As a result, there are numerous Sacred Lotus plants germinated from fruits accurately ^{14}C-dated at 150, 200, 450, 600, 700 years now thriving and flowering at the UCLA Mildred Mathias Botanical Gardens, the Wuhan Botanical Garden in southeastern China, and on her lab bench and in huge pots at her and Bill's home, where she continues to study their growth.

This work matters. Lotus is a cultivated food-source throughout Southeast Asia. But the average shelf-life of viable crop-plant seeds (e.g., corn, rice, wheat) is only some 10–20 years, not hundreds or more than a thousand years. Such longevity is genetically determined. To investigate this in lotus, after several years and numerous tries, Jane found a receptive expert in plant genomics – Professor Ray Ming at the University of Illinois-Urbana – with whom she and some 70 other colleagues have done the science required to decipher the complete gene sequence of *Nelumbo nucifera*. The task now underway is to identify the genes and gene complexes responsible for the remarkable long-lived viability of lotus fruit and to ultimately transfer that longevity-encoding information into other crop plants, perhaps into endangered plant and animal species and, possibly, even into humans. Once accomplished – studies with which Jane is currently having great fun – this work could have a major impact on rural economies and relief from famine throughout the world, and it might even help offset the disastrous biotic effects of global warming and increase the quality of long-lived human life.

Jane and Bill both work seven days a week; are truly interested in each other's areas of study; and respect, appreciate, and are inspired the other's values, goals, and work ethic. Indeed, they interact so well that they are rather commonly asked "How long have you two been married?" (Bill's private, mildly humorous response – which he has not yet ever actually used – would currently be "68 years. I was married for 13 years, she for 17 years, and the two of us to each other for another 38!")

Obviously, for Bill Schopf, the decade of the 1970s was remarkable, a period when time and time again all fell in place – Lunar sample studies; the Oparin visit to his lab; the Waterman Award; multiple research sojourns to India, Russia, China; and a new amazing soul-mate. More importantly for the development of the then fledgling paradigm-shifting development of Precambrian Paleobiology, these adventures resulted in finds of the first Precambrian microbial assemblages of India, the former Soviet Union, and China, providing the needed major impetus to begin to move this science toward its present-day international, interdisciplinary success and the basis that set the international astrobiological community on its quest to search for such evidence on Mars. *Wow, what a wonderfully rewarding decade of continuous good luck!*

7 The 1980s – The PPRG Defines the Field

7.1 PRECAMBRIAN PALEOBIOLOGY RESEARCH GROUP

Receipt of the Waterman Award in 1977 handed Bill Schopf an unexpected opportunity. But with that opportunity, he realized, came special responsibility. How best could this entirely unforeseen windfall of funds be used?

By that time, Bill was a fully confirmed interdisciplinary internationalist and had met most of the major figures in the field and many of their students. As a result, he had mused about putting together a "pipe-dream team" of Precambrian workers who would come to his lab at the University of California Los Angeles (UCLA), pool their knowledge and rock samples, and work together for perhaps a year. Because of his background as a musician, he pictured this to be a science-version of a "Dixieland Band," one playing a trumpet, another booming-forth on a trombone, a third tickling a bass fiddle, a pianist stroking the keys, a drummer beating the skins. In other words, it seemed natural to him that such a science team would need diverse expertise – in Geology, Microbiology, Biochemistry, Evolutionary Biology, Paleontology, Atmospheric Evolution, and Isotopic and Biogeochemistry – his hope being that the differences in background within the group could be overcome by each member teaching the others about their specialty. And his notion was that the team-members should be relatively young, his aim being to set the field on a new course not constrained by accepted dogma and the predilections of their elders, an approach that would lead usefully into the future.

Bill's four goals were simple, intended to mold a "new" (in fact, first full-blown) version of Precambrian Paleobiology that would be: (1) interdisciplinary, (2) international, (3) free of strife (a problem he had seen at Harvard that he wanted to avoid), and (4) a model for subsequent Waterman awardees – given that he was only the second recipient of this virtually unprecedented national distinction – that he hoped would illustrate how this unique opportunity might be used most effectively.

Bill wrote up his ideas and sent them off to two friends: John Hayes (Figure 7.1), an organic geochemist Massachusetts Institute of Technology (MIT) student whom he had known since graduate school; and Malcolm Walter (Figure 7.2), Australia's leading young Precambrian stromatolite expert, whom Bill had met when he visited his professor, Martin Glaessner, in Adelaide. Both were (quite reasonably) skeptical, but both trusted Bill (for which he was and is deeply appreciative). They both agreed that his plan was worth a try.

Though interdisciplinary science, the hallmark of Precambrian Paleobiology, is now widely accepted (for example, in Astrobiology, Molecular Genomics, Human Biology and numerous other fields), in the 1970s it was not. Not surprisingly, this aspect of Bill's plan – coupled with his emphasis on "Young Turks"– presented

FIGURE 7.1 John Michael Hayes and plant biologist Jane Shen Schopf (the author's wife).

FIGURE 7.2 Malcolm R. Walter.

problems. Once he did the math, it was clear that the Waterman $150K would not be enough to fund the venture. He therefore prepared a grant proposal to NASA's Exobiology Program requesting matching funds. The two great US giants of the science were the reviewers. Both turned it down: Barghoorn, because he thought "good science is done by individuals, not by groups"; and Cloud, because the plan "did not include senior workers experienced in the field." Once again, Bill lucked out. Despite these negative assessments, the NASA program officer (Dick Young) funded the project.

OK, Bill's notion was both innovative and risky. But as the old adage has it, "nothing ventured, nothing gained." Just as he had envisioned, the team (initially, 19 scientists from four countries) – calling itself the Precambrian Paleobiology Research Group (PPRG) – came to UCLA with their families in tow, the PPRGers working together in Bill's lab for 14 months. This was enormously great fun! Every Wednesday evening, they gathered for pizza and beer and heard a lecture about each other's work.

When in early 1980 Bill and his fellow delegate on the 1978 trip to China, plant-biologist Jane Shen-Miller, finally found time to become married – before the state-required medical permissions had expired – they staged a "special" Wednesday night PPRG gathering, replete with a wedding cake and wine, their version of a wedding reception. The very next day, origin-of-lifer Jim Ferris from Rensselaer Polytechnic Institute came to visit Bill and asked him: "How are things going?" Bill replied: "Quite well – I got married yesterday." (Ferris came close to fainting!)

In 1983, the PPRG published its first results, *Earth's Earliest Biosphere, Its Origin and Evolution* (Figure 7.3), spanning the Hadean and Archean, the earliest two billion years of Earth history. A few years later, with funding from NASA, the National Science Foundation (NSF), and the National Geographic Society, Bill put together a second PPRG, now including 40 scientists from eight countries, to focus on the history of life and of the Earth during the Proterozoic, the more recent two billion years of Precambrian time. In 1992, they produced a second major volume, *The Proterozoic Biosphere, A Multidisciplinary Study* (Figure 7.4). Given the spans of geological time the volumes sequentially covered, friends in the field have jokingly referred to them as the "Old Testament" and the "New Testament" – and as Bill's colleagues and he had hoped, they had enormous impact worldwide, serving as "career starters" for many young scientists who have since gone on to make important new finds and major contributions.

The two PPRG projects were monumental efforts to which Bill devoted a total of 14 years – seven for the first PPRG (1977–1983; Figure 7.5) and another seven for the second (1985–1992) – organizing the groups, preparing grant proposals, curating the pooled rock samples, hosting each group for a 14-month-long stint of lab work at UCLA (and their weekly pizza-beer confabs), and doing science, writing up his contributions, and editing the resulting two massive volumes. Not only did the PPRGers pool specimens of their relevant earlier collected rock specimens but members of the team, typically in 8–12-member groups (led by a PPRG member well acquainted with the region of interest) also carried out additional field work and research in western and central Australia, southern and Arctic Canada, eastern China, northern Mexico, South Africa, the USSR, Europe, and the United States. The rocks collected

FIGURE 7.3 The 1977–1983 PPRG volume *Earth's Earliest Biosphere.*

FIGURE 7.4 The 1985–1992 PPRG volume *The Proterozoic Biosphere.*

FIGURE 7.5 The 1977–1983 PPRG (left to right, front to back) Manfred Schidlowski, Sherwood Chang, Bill Schopf, David Des Marais, Kim Wedeking, Jim Walker, Malcolm Walter, John Hayes, Ian Kaplan, Howard Gest, David Chapman, Keith Kvenvolden, Gary Strathearn, George Claypool, David Stevenson, Gary Ernst, Kase Klein, Stanley Miller.

during this work were then added to the previously pooled samples (all of which remain at UCLA, having been used over subsequent years by scientists across the globe).

As a result of this work, the PPRG amassed a collection of more than 2,000 samples from hundreds of geologic units; carried out analyses of the carbon and sulfur content and isotopic compositions of more than 300 of these units; and evaluated a total of 525 units for microfossils, of which 400 contained bona fide microbial fossils, many new to science.

Moreover, the PPRG was broadly interdisciplinary – including experts schooled in atmospheric evolution, comparative planetology, prebiotic organic syntheses and the origin of life; field geology, mineralogy, paleontology, sedimentology, and plate tectonics; organic chemistry, biochemistry, biomarker and isotopic geochemistry; and organismal and molecular biology, genetics, microbiology, and evolutionary biology. Because they worked together day-in, day-out, each garnered the respect of the others. And they all learned from one another (an aspect of the venture fueled by their Wednesday evening three- to four-hour pizza-beer confabs – lots of questions, lots of answers, and lots of new ideas and new knowledge – to the entire group, an enormously educational experience).

For those interested in how science actually works and the roots of the paradigm shift to which this team so markedly contributed, they need to look no farther than the goals, structure, and camaraderie of the PPRG. The friendship, mutual respect, and common values of the group came to the fore – despite the incessant interruptions and probing queries posed by other PPRGers during each and every Wednesday evening presentation. All knew that the questioners were serious fellow scientists seeking to become educated about the presenter's area of expertise. The discussions were honest, open, often jovial, and never acrimonious. That is the way science should be. And that is the explanation for why the field of Precambrian Paleobiology today is so markedly, so wonderfully, interdisciplinary, international, and effective.

The products of this work – the two PPRG volumes together spanning the entire Precambrian from the formation of the planet 4,500 million years ago to the rise of metazoans about half-a-billion years ago and extending the known record of life seven-fold from what was known in Bill's college days – each received the US National Scholarly Book Prize in its year of publication (Figures 7.6 and 7.7). Together they have set Precambrian Paleobiology on solid footing and shaped the course of the science. Moreover, they have provided the crucially required sound basis for the solution of Darwin's dilemma, the problem Bill set out to solve years earlier when he was a 2nd-year know-nothing college student. (As you might imagine, in retrospect he is wonderfully pleased by their success.)

The two PPRG ventures also helped to mold the then-nascent field of Astrobiology, as NASA deputy director Jerry Soffen explained to Bill when he visited his home to quiz him about the success of the PPRG while Soffen was struggling to establish the NASA Astrobiology Institute. PPRG's influence on that science is palpable, numbering among its alumni more than two dozen notable contributors to the now well-established field of Astrobiology. Indeed, the Mars 2020 mission designed to cache rocks for delivery to Earth, intended to include promising repositories of evidence of ancient Martian microbial life, is yet another product of PPRG's influence, one of the NASA overseers of the project telling him that "Bill, it is all based on your work" (a no doubt well-intended bit of hyperbole – prompted by Bill's discovery of fossil microbes preserved in gypsum, a rock type widespread on Mars – that would more correctly be worded "the work of the PPRG and many others worldwide"). And every prime member of the PPRG has grown over the decades following their endeavors to become an international leader in the field, garnering honorific memberships, medals, awards, and well-earned widespread appreciation.

Of Bill's four goals for the PPRG, two were met: Precambrian Paleobiology is now and for the foreseeable future will continue to be both interdisciplinary and international. His other two goals failed, largely he imagines because of what he terms the "me first" gene that is deep-seated in us human social primates. In any case, he has come to realize that his third goal, an absence of strife in the field – not among PPRGers but in the field as a whole – may not be attainable (though he still wishes it were a universal hallmark of science). And his fourth goal, the notion that the Waterman funds might most effectively be used for field-promoting "group-first" endeavors, has evidently been adopted by only one other of the now 42 awardees

FIGURE 7.6 1983 PPRG Award (Physical Sciences), Association of American Publishers.

FIGURE 7.7 1992 PPRG Award (Geography and Earth Science) Association of American Publishers.

(a psychologist at the University of Washington, the 2018 recipient). Don't fault the other awardees – they all are in fields markedly different from Bill's and have no reason to have even the slightest inkling as to how he used his prize money. Indeed, it has become common practice for the recipients to use such prizes to "buy their contract" from their home institution, giving them four or five years of "freedom" from institutional obligations.

In total, the two PPRG ventures were composed of 50 scientists (13 now deceased) from 30 academic institutions and eight countries: Australia: Malcolm Walter (Univ. New South Wales); Canada: Hans Hofmann (Univ. Montreal) and Ján Veizer (Univ. Ottawa); Denmark: Bo Barker Jørgensen (Aarhus Univ.); Germany: Udo Matzigkeit and Manfred Schidlowski (Max Planck Inst. Mainz) and Harald Strauss (Univ. Münster); Sweden: Stefan Bengtson and Gonzalo Vidal (Uppsala Univ.); South Africa: Nic Beukes (Univ. Johannesburg); Russia: Mikhail Fedonkin (Geol. Inst. RAS Moscow); and the United States: Dick Castenholz (Univ. Oregon); Sherwood Chang and David Des Marais (NASA Ames); David Chapman, George Claypool, Ray Ingersoll, Ian Kaplan, Bruce Runnegar, Bill Schopf, Gary Strathearn, and Toby Moore (UCLA); Sam Epstein, John Grotzinger, Joe Kirschvinck, and David Stevenson (CalTech); Gary Ernst and Don Lowe (Stanford Univ.); Jack Farmer (Arizona State Univ.); Walter Fitch (Univ. CA, Irvine); Bob Garrels and Dick Holland (Harvard Univ.); Howard Guest, John Hayes, and Kim Wedeking (Indiana Univ.); Jim Kasting and Lee Kump (Penn State Univ.); Kase Klein (Univ. New Mexico); Bob Horodyski (Tulane Univ.); Jere Lipps and Don DePaolo (Univ. CA, Berkeley); Carol Mankiewicz and Carl Mendelson (Beloit College); Stanley Miller (Univ. CA, San Diego); Beverly Pierson (Univ. Puget Sound); Jack Sepkoski (Univ. Chicago); Roger Summons (MIT); Ken Towe (Smithsonian Inst., Washington, D.C.); Jim Walker (Univ. Michigan); and David Ward (Univ. Montana).

For all involved, participation in this bold experiment was a landmark experience in their careers, to which each contributed expertise and myriad accomplishments. In July 2014, two decades after the publication of the second major PPRG volume, Bill put together a three-day reunion for the group at UCLA, during which PPRGers gave scientific presentations which were followed by an outing to local museums (Pleistocene fossils at the La Brea Tar Pits and the Space Shuttle at the L.A.'s Science Museum), the festivities being topped off by Bill's wife Jane orchestrating a gala dinner at their home.

Thirty-two (of the 35) PPRGers then still alive, mostly by then retired, managed to attend. Of these, 17 were accompanied by their spouses, their life-long compatriots who were no doubt curious as to how and why their marriage partners had managed to have so much fun with the PPRG gang. As Bill toted up the awards and prizes to prepare a program for that gathering, he discovered, to his immense pleasure, that among these 35 PPRGers, 13 were members of their National Academies of Science (Australia, Canada, Russia, Sweden, United States) or its UK equivalent, the Royal Society of London; that the 35 PPRGers had received 33 national/international medals, nine teaching awards, and five honorary doctorates; and that five were recipients of highest accolade in the field, the US National Academy of Sciences' Charles Doolitttle Walcott Medal. *Hooray for the PPRG*!

Following this meeting, Bill sent the following short note to the group:

Wednesday 30 July 2014

Dear PPRGers:
Big thanks for coming to the PPRG Reunion!

For Jane and me it was plenty good fun (or, as my student, Ms. Amanda, told me, it was "unlike anything I have ever seen" ... a "blast"). She is right – the pleasure, the sheer joy we had being with one another is unparalleled, unlike anything I have ever before experienced. As Jim Walker would say, "it was a good show," or as Kase Klein, and Malcolm Walter, and John Hayes and a whole bunch of others would say, "we done OK and maybe a bit more."

Think of it. There are only 35 of us still extant and 32 of us showed up (with a 33rd, Misha Fedonkin, very much wishing that he too could be with us). And those 32 were accompanied by 17 spouses. As far as I am concerned, that is Great Stuff! I don't know about you folks, but for me I learned long ago that without a supportive spouse, my life, my science, would be a terrible mess. The support and love that so many "better halves" (for me, the "better 2/3rds") showed us by their presence is truly wonderful.

The bottom line: Unearthing of the crucial data required to establish the paradigm shift from the previously unknown to the now known early record of life – in Kuhnian phraseology, the "articulation" of the paradigm – has been an international interdisciplinary group effort, the decades-long work of many scientists from many counties, experts in fields that others may have imagined to be unrelated areas of science. Yes, Bill Schopf was instrumental and played his part. But this was a preplanned, carefully focused *group endeavor*. To truly comprehend how this paradigm-changing science has advanced, it should be understood that it is the PPRG as a whole – not any single person – that deserves the kudos for moving this field ahead.

7.2 AMERICAN PHILOSOPHICAL SOCIETY

In 1985, Bill was elected to the American Philosophical Society (APS). Founded in Philadelphia by Benjamin Franklin in 1743, the APS is the oldest and most distinguished scholarly honorific society in the United States, its membership of some 825 filled with exceedingly eminent (typically, octogenarian) luminaries. It is broad and wonderfully deep, including scientists, historians, philosophers, artists, musicians, justices of the US Supreme Court, former US presidents and senators, a covey of Nobel laureates, and many others. As his fellow paleontologist and APS member Norman Newell explained to him, "If the National Academy of Sciences can be thought of as the All-Star Team, the APS is the Hall of Fame."

Bill was elected to the Society at the young age of 44. Evidently because of his relative youth, he was soon asked to present a 20-minute talk at the Annual Spring Meeting, at that time convened in Philosophical Hall, the handsome building directly adjacent to Independence Hall where both the Declaration of Independence and the US Constitution had been put in place. Jane and he arrived in Philadelphia the day before the meeting, and he spent much of that afternoon perusing the holdings of the Society's library, across the street from Philosophical Hall, which had amassed a remarkable collection of Darwin's original papers as well as those of many other APS-member notables in his areas of interest. And because for the week immediately preceding their trip to Philadelphia the two of them had been at the University of North Carolina in Chapel Hill, where Bill delivered the annual Paul Mangelsdorf Lectures, rather than attending the Society's evening concert the evening before his maiden APS talk, he remained at the hotel rearranging his slides and preparing his lecture.

The lecture room in Philosophical Hall, constructed in 1785 and designed for oratory, is broad and shallow, extending left to right relative to a small stage and lecture podium at its front, in this respect differing markedly from its modern counterparts, which are designed for visual presentations and are thus narrower and deeper, the audience seated from front to back facing the projection screen. Other than having improved seating and an up-to-date audio system, the lecture hall had remained basically unchanged for more than 200 years, fronted by the same wooden podium from which Franklin, George Washington, Thomas Jefferson, John Adams, Alexander Hamilton, Thomas Paine, James Madison – all members of the APS – had addressed the Society.

Bill's turn in the morning's program arrived. He ascended the small stage and asked for his first slide (in those days, 35mm slides mounted in a carousel). The slide appeared … it was his, but it was not the first in the lecture. He then asked for the second, then the third. They too were his, but again, they were the wrong slides! Thinking that the projectionist had dropped the carousel and hastily rearranged the slides, he walked the short distance to the projector to sort it out. He then realized that "Oh my! I brought the wrong carousel!" His carefully crafted lecture was in the other carousel still sitting on the desk in his hotel room!

Emergency time! What was he to do? As he walked back to the podium, three options came to mind: (1) retrieve the correct carousel from the hotel – but that was a 20-minute walk away, and by the time he returned, his place in the program would have passed; (2) cancel the lecture, apologize to the Society, and get off the stage – doable, but him being a newcomer to this august group of "Hall-of-Famers," this would have been terrifically embarrassing; and (3) suck up his gut, clench his teeth, and weather this ordeal by making up a new lecture on the spot. During the brief stroll back to the podium, he muddled through the options and told himself, "These folks elected you to this Society and invited you to talk – meet your obligations! – you simply *must* perform!"

As adrenalin surged though his veins he didn't outwardly panic – but in fact he was frightened to the core – this being worse, far worse, than the time he had been stranded alone when his legs became dysfunctional as he was climbing a nearly vertical thousand-foot-high rock wall in Western Australia. He had worn his (one and only) business suit, but inwardly he felt completely naked, stripped to his bare skin. What should he talk about, and how should his presentation be structured?

Bill then embarked on the oddest, strangest, scariest episode of his life. Because he knew the history of the century-long search for the "missing" Precambrian fossil record, and because he had discovered the previous afternoon that virtually every one of the "Biggies" in the field had been an APS member, he decided to tell the audience about the origin and history of this reputedly "greatest unsolved problem in Natural Science." This, he thought, was a story they might enjoy, since he could peg it as showing the role of the APS in finally leading to its solution.

So far, so good. But what would he say and how could the story be properly presented in the allotted time? Bill still remembers, as though it happened only yesterday, the terrifying feeling of trying to sort this out. He said to himself, "Brain, right side, compose the lecture" – not too arduous a task, since he could handle the subject chronologically. But he was seriously worried about getting the words out – "Brain, left side, tell the mouth to speak" – because he had an awful fear that his words would emerge as an unintelligible jumble of nonsensical blathering babble.

He was consumed by these thoughts and fears for the 20 seconds or so it had taken him to resume his position at the podium. His cerebral hemispheres then got to work. He didn't explain the situation except to say, "OK, we'll do this talk without slides."

Bill began the lecture. Thankfully, the words came out. He began with APS-member Charles Darwin, who in 1859 had stated the problem of the "missing" fossil record of early life, and then moved to the work of Sir William E. Logan, director of the Geological Survey of Canada, to whom in 1858 a collector had given specimens collected from Precambrian rocks along the Ottawa River that Logan thought might be Precambrian fossils. The next in line was the staunch Calvinist Sir John William Dawson, principal of McGill University, to whom Logan gave the specimens for study and who then misinterpreted them to be fossils shells of giant protozoans and named them *Eozoön Canadense* ("The Dawn Animal of Canada"). Although the structures were shown by the 1870s to be assuredly non-biologic, Dawson persisted in regarding them as bona fide fossils, and – noting that "There is no link whatever to connect *Eozoön* with younger fossils" – he interpreted them to be a product of divine special creation that thus provided "damaging negative testimony against evolution, [a theory] incapable of proof and contrary to fact [promoted by] mere dreamers, having no scientific basis for their dogmas."

Next, Bill outlined the contributions of Charles Doolittle Walcott, a former president of the APS who, during his harrowing geological excursions down the Colorado River through the Grand Canyon in the1880s and 1890s, had discovered what are now acknowledged to be the first bona fide Precambrian fossils known to science – only to have them discounted for the following 60 years. This Bill then followed by recounting the contributions of the two APS-member pioneering masters of the modern science, Elso S. Barghoorn (Bill's graduate professor) and Preston E. Cloud, Jr. (who was in the audience); and concluded with his own contributions.

The yellow light on the podium flashed on, telling Bill that he had 60 seconds remaining. He summarized the talk in three main points – items that he cannot now recall – and he ended just as the red light came on. Great, he had finished on time! The reaction of the audience was amazing – applause, enthusiastic cheers, some of the octogenarians actually rising from their chairs – appreciation that no one has ever witnessed before or since at a meeting of this august and typically rather staid Society. Wow!

In the question–answer session after the talk, Bill was asked by his former Harvard organic chemistry professor, "What was the problem with the slides?" Bill replied, "I brought the wrong carousel," owning up to his grievous error. A coffee break directly followed his talk, and the group went upstairs to refresh and relax. Bill was still reeling, his hand shaking as he tried to get a coffee cup to his lips. Pres Cloud then came over and said, "Well, Bill, you'll certainly get a lot of mileage out of that!" Cloud smiled, but he was more than a little miffed (his annoyance being predictable. Cloud had never much liked Bill ever since he turned him down to instead do graduate work with Barghoorn.)

The upside of this stressful episode was that Bill's dear wife Jane later told him "that is the best lecture I've ever heard you give" – an enormous compliment, given that she had been present at a great number of his previous (properly prepared) harangues. And as it turned out, she, too, became a beneficiary of the episode. Bill and his wife were then APS neophytes, mere "youngsters" in the Society, and she had been understandably more-or-less shied away from by a number of the "old-timers," including the wife of Harvard Astronomer Fred Whipple (a 1927 UCLA Mathematics major who in the early 1950s was the first to identify comets as being "dirty snowballs"). After Bill's talk, however, Babette Whipple (whom Fred had met during his UCLA days) befriended Jane and graciously complimented her on his presentation. And later that day, another old-timer who previously had been rather aloof, Margo Hamp (wife of the famed authority on Indo-European and historical linguistics, University of Chicago Professor Eric Hamp) presented Jane a handsome red-coral necklace (a much appreciated replacement of one that Jane had much valued, stolen from her checked baggage upon arrival at Moscow's Sheremetyevo Airport during a recent trip to the USSR). My Goodness, this time Jane and Bill both were plenty lucky!

As a final little aside to Bill's now more than three decades' association with the APS, he has been invited to give a talk at the April 2018 celebratory 275th anniversary of the Society. Why him? It's simple. He studies fossils and the history of life. And so did US vice-president and then president Thomas Jefferson who while so serving was also President of the APS from 1797 to 1815. Moreover, and although it is not general knowledge, Jefferson was the founder of Vertebrate Paleontology in North America, a result of a presentation he made at the 1797 APS meeting in which he unveiled *Megalonyx*, a giant ground sloth excavated a year earlier in what was then known as western Virginia. Jefferson's paleontological prowess is remembered even today, the official "state fossils" both of Virginia (*Chesapecten jeffersonius*, a Miocene-age extinct scallop) and West Virginia (*Megalonyx jeffersonii*) being named in his honor. Bill's forthcoming 2018 talk to the APS is entitled "Pioneering APS Paleontologists Jefferson, Darwin, and Walcott – the Known History of Life, Then and Now." Still – and though Bill will no doubt try to do his best – however this talk is received, it will pale in comparison with his first terrifying presentation to this gathering of "Hall-of-Famers." [Added note: The talk went well; Washington, D.C. Federal Court Judge David Tatel graciously opined that it was "beautiful"; and importantly to Bill, wife Jane was pleased.]

7.3 WHAT CAN BE LEARNED FROM THIS?

On occasion, Bill has recounted a shortened version of his initial APS plight to his graduate students. Why? For him, it was a dreadful (if obviously memorable) episode, a situation that he would never wish on his "greatest enemy" – even if he had one, which he does not. But it was also a meaningful object-lesson that illustrated two simple truths: (1) You learn more from your mistakes than you do from your successes; and (2) you have no idea how you will cope with a crisis until you are actually confronted with one.

Had Bill been asked beforehand whether he could handle such a situation, he would have replied: "Are you kidding? No way!" But when it actually occurred, he somehow found a way to cope. Some, one might imagine, would regard his response to this situation as an act of courage – though he personally has always been impressed by the fine lines between courage, stupidity, and fit-in/follow the crowd "group-think" (a set of contrasts illustrated by wondering about the GIs charging out of foxholes into a barrage of bullets during World War I). Be that as it may, this episode illustrates yet a third basic truth, namely, that "nothing ventured, nothing gained" (though, of course, the reverse is also true: "nothing ventured, nothing lost").

If Bill could manage this unforeseen potential disaster, he imagines that the students can too, his object being to help the students to believe in themselves, to know that down-deep they have what it takes to weather even the most disastrous storms. And he'd like them also to realize that although risks in life, as in science, carry the potential for ruinous consequences, they also present an opportunity for success. (The key, Bill supposes, is to sort through the options and pick and choose between when to "play your cards" and when to "fold" – and for this, sorry to say, he has no foolproof recipe.)

7.4 UCLA'S HONORS PROGRAM AND THE CENTER FOR THE STUDY OF EVOLUTION AND THE ORIGIN OF LIFE

For several years in the early 1980s, Bill served as dean of the Division of Honors of UCLA's College of Letters and Science. This program was designed to encourage absolute excellence among the top 10% of the university's 25,000 undergraduate students, his goal being to use this position to provide a high-quality "Oberlin-like" experience for these young potential leaders.

Bill expanded the Departmental Honors Program (and its requirement for a Senior Honors thesis) to include all departments throughout the College, and, using funds donated by biographical novelist Irving Stone (author of *Lust for Life*, *The Agony and the Ecstasy*, and many other works) he then established the College Honors Commons, a well-equipped meeting area intended for lectures, concerts, one-act plays, poetry and literature readings, and topical discussion sessions. Following these successes, he initiated the Undergraduate Student Research Program, a venture appropriate for a major research university such as UCLA that Bill's wife Jane urged him to pursue. The UCLA program, in place now for decades, has been highly effective campus-wide, providing funds and course-credit that enables Honors students to carry out research projects during both the academic year and the summer recess under the tutelage of UCLA professors.

Bill much enjoyed his dean's job and its responsibilities – he had vision, he knew how he wanted to proceed, and his notions were paying off. But he was soon confronted with a conundrum. By that stage in his career, his home department had evolved from having six colleagues in his areas of interest to but three: Clarence Hall, who by then had been appointed dean of the Division of Physical Sciences (a position that he had no intention of vacating); Helen Tappan Loeblich (a world-leading micropaleontologist); and himself. When Professor Loeblich elected to retire early, Bill had no choice but to return to his department.

Soon thereafter, UCLA gave Bill yet another signal opportunity, namely, to found and serve as director of the interdepartmental-interdisciplinary Center for the Study of Evolution and the Origin of Life (CSEOL). Clearly, at that time UCLA – still now ranked in the top 10 universities in the world – was willing to "take chances." This was a novel idea, and the University trusted him to deliver. He never got any teaching relief or extra salary from this – he never asked and he didn't care – but he did deliver.

Meeting weekly in the Center's library, adjacent to Bill's office, 27 times during each academic year, CSEOL hosted weekly three-hour dinner-discussion sessions of WEEG (the Wednesday Evolution Group) attended by 40–55 faculty and invited graduate students from diverse UCLA academic departments. And everyone played the game by Bill's rules which, though initially alien to many, had worked well with the PPRG – namely, if you don't understand something, interrupt and ask a question, preferably at the end of a paragraph – an anathema to many who were used to the standard departmental seminar format of having a presentation being followed by one or two perfunctory inquiries.

WEEG participants were drawn not only from the Life and Physical Sciences but also from the Social Sciences (e.g., Anthropology, Geography, Economics); Humanities (e.g., English, Philosophy); the UCLA Medical School; and other universities (e.g., CalTech, U.C. Riverside, U.C. San Diego, the University of Southern California, and various State University of California campuses). CSEOL also sponsored numerous evolution-related courses in various departments and each year provided fellowship stipends to six especially promising incoming graduate students in any of 8–12 UCLA academic departments.

During this period, CSEOL also convened 17 annual all-day, free-of-charge, open-to-the-public all-campus symposia featuring such luminaries as Nobel Laureates Tom Czech, Christian De Duve, Wally Gilbert, Manfred Eigen, and Jack Szostak; NASA administrator Dan Goldin; Pulitzer Prize-winning geographer Jarrod Diamond; *Nature* senior editor Henry Gee; London's Natural History Museum paleobiologists Richard Fortey, Paul Taylor, and John Richardson; origin of life scientists Stanley Miller and Leslie Orgel; paleoanthropologists Phillip Tobias, Alan Walker, and Chris Stringer; psychologist and linguist Steven Pinker; evolutionary biologists Ernst Mayr and Stephen Jay Gould; molecular phylogenists Walter Fitch and Tom Jukes; astronomers Carl Sagan, Frank Drake, and Ned Wright; and famed novelist Irving Stone. The symposia were highly successful, each year filling an auditorium with more than 550 attendees, and each resulted in a published educated layperson-level book.

CSEOL's weekly dinners were wonderful – open, honest, unadulterated intellectual fun – regarded by many of the participants (then and even to this day) as their "very best experience at UCLA." Just like PPRG, all enjoyed the opportunity to learn from others.

For Bill, however, CSEOL was also arduous. For a considerable period, he had cajoled others in the group to prepare the weekly dinner … a huge help! But over the years they grew tired of that ritual, and because he no longer had an administrative assistant, keeping WEEG alive fell to him. Bill didn't mind organizing the weekly talks and the yearly symposia – Good Fun! But to arrange the dinners he had to go off-campus two or three times a week and then, back in the CSEOL library, to arrange the chairs, clean the tables, put out the utensils, organize the drinks, put the dinner in place, and so on. After 25 years, he finally decided that "enough is enough." Bill wrote to the group what he intended to be a nice letter of closure titled "All Good Things Must Come to an End." Many graciously wrote back and, though dismayed, seemed to understand; but a few have been miffed ever since. (Take-home lesson? Try your best. Do your best. But even your best efforts are unlikely to please everyone!)

7.5 TEACHING AND STUDENTS AT UCLA

On the bases of three broad criteria – research, service to the community, and teaching – UCLA faculty are formally evaluated regularly by their department colleagues, every three years for assistant and associate professors and every four years for full professors. In Bill's early days in the department, his colleagues regarded this process with considerable seriousness, both "decelerations" (for mediocre performance) and "accelerations" (for perceived excellence) being not uncommon. As a result of such in-depth evaluation, in the 1970s his UCLA department was ranked third in the United States (a ranking he particularly now recalls because the department of his older brother, Tom, at the University of Chicago, was ranked fifth – a comparative scoring that rankled Tom no-end). More recently, however, as the department expanded both in size and breadth, such faculty evaluations have become more perfunctory (except for promotions, nominally after a six-year "trial period," from assistant professor to the tenured position of associate professor).

Despite these changes in its perception and execution, throughout Bill's career, he has always regarded the UCLA evaluation system to be an excellent process. It is fair, egalitarian – everyone knows the rules; everyone knows the expectations. Members of the faculty have an opportunity to review a candidate's dossier and, thereby, to become acquainted with the candidate's field and recent activities and to personally assess whether the candidate is or is not "measuring up." To Bill's mind, the alternative – the "null hypothesis" by which in some universities such decisions are made by a permanent department head (rather than having the faculty led by a limited-term department chair) – is far less appealing. Indeed, this alternative system tends to result in authoritative "my way or the highway" governance; secretive collusion ("you scratch my back, I'll scratch yours"); divisive balkanization of subdiscipline-defined factions of a department faculty; and the encouragement of "blackmail" of the department and the university ("Gimme the promotion or I'll take a job offer elsewhere"). In short, the alternative process fosters egocentric me-firstism, the antithesis of the group-first, society-first attitude that is central and crucial to excellence in academia and to the betterment of the broader society.

OK ...Bill has prospered under the UCLA system and, thus, cannot be said to be an "objective evaluator." So be it. To him, however, it's the best, most open and honest system he knows – it fits his values. And he has tried his best, now over five decades, to measure up in all three categories. His research contributions have made a difference, and his "community service" as dean of honors and director of CSEOL has been acceptable – but what about his teaching?

As noted above, as a neophyte assistant professor, he was taught how to teach by fellow Oberlin graduate N. Gary Lane, whose mantra was "the students come first." Gary and he shared similar backgrounds and the same values, and throughout Bill's career, he has tried to follow Lane's dictum.

Due to the efforts of Gary Lane, Clarence Hall, and the students in a course that Bill shared with them, in 1977 he was recipient of the university's annual campus-wide Distinguished Teaching Award, the first in the history of his department. Since that time, he has continued to try his best to help the students but by the late 1980s he had become disgruntled by the reluctance of some students to visit him during the standard two-hour-per-week "Office Hours" sessions. He wanted to help them, especially if they were having difficulties with his class, but they had innumerable excuses: "Your Office Hours conflict with another class" – "I had to work at my job" – "I have other [unspecified] obligations" – "I overslept/missed the bus/had a flat tire on the way to school." (At the same time, some of their explanations were to Bill heart-rending, serious problems that hit him to his core.)

Given this, Bill decided to "remove the rug from beneath the students' feet" by expanding his availability in his Tuesday and Thursday predominantly Freshman-Sophomore General Education class from the standard two hours weekly to 12 hours per week, eight o'clock in the morning until noon, every Monday, Wednesday, and Friday. (Some years later, he decreased this to 9am to noon – thus shortening his availability to nine hours per week – because he had come to realize that students never ever showed up in the 8 to 9am period and that the freed hour would give him a chance to deal with his daily influx of email messages.)

Bill has never mentioned any of this to his department colleagues – not his style. But a few discovered his ploy, either from students or from their periodic review of his dossier, and at a department faculty meeting he was chastised by one of his colleagues: "Don't pay any attention to Schopf's teaching record – he panders to his students." A non-academic might find this odd, even inappropriate, but to Bill it was no great affront and, actually, seemed easily understandable. In the students' ratings of his performance, they had regarded him highly, perhaps (one can only surmise) somewhat better than his outspoken colleague. So it goes. It didn't matter to Bill – he was simply doing what he thought was right.

Over the years, Bill has learned an enormous amount from his interactions with students. When they come to see him, his more or less standard approach is to ask them how they are doing in their other classes (his notion being that if they are having problems with his course, they are likely to being having difficulties with their others) and to ask them about their family background, their sisters, brothers, and parents (to him, a useful indicator of their experience and mind-set). Virtually all of the students open up. They trust him; they know that he genuinely cares. Why are they so trustful? It's simple. Every single undergraduate at UCLA has the ability to excel – otherwise they would not be at UCLA as a member of an entry class of ~5,000 out of an applicant pool of nearly 120,000 (the largest in the country). Their difficulties stem, time and again, from being on their own for the first time in their lives and not knowing how to effectively use their time. Bill gives them tips, he tries to help them, and it often pays off.

Not at all surprisingly, the society-based changes in the evolution of the fabric of academic departments over the past half-century could, one imagines, have been predicted years earlier from the evolving behavior of the students. In fact, virtually all of the major changes that have occurred over this period were previewed by college student activist movements – everything from the idealistic "improve the world" focus of the 1960s, to the promotion of racial integration, then women's rights, then the acceptance of homosexuality and same-sex marriages – perhaps now, even "gun control." It seems likely that these and other less monumental societal changes might well have been seen on the horizon by those studying the attitudes of college students, and most recently those of high-school students as well.

Given this and Bill's frequent interactions with present-day students, he has some concerns. As you will have gathered, he is an optimist, not even close to being a "doom and gloomer." At the same time, he tries to be realistic and cannot help but compare the freedom he had as a youth with what the current students tell him about their upbringing. As a simple example, when Bill was young, it was not uncommon for him to ride his bicycle several miles each day – to deliver newspapers, to attend the practice sessions of his sports teams, to build forts with his pals on the banks of the nearby river, or to visit his girlfriends – and his parents expressed no concern. But according to today's students, things have changed, their parents prohibiting them from riding their bicycles more than a block (or less) from their homes and certainly not to some distant soccer practice or ballet class, their "helicopter parents" serving as chauffeur. This, coupled with "social advancement," the practice of promoting students from one grade to the next regardless of their mastery of required subject matter – told to Bill by high-school teachers to be a common occurrence "so as not to hurt the students' feelings" – and the practice of giving everyone on a youth sports-team a trophy for simply showing up, not for performance, seems likely to have eroded the initiative and self-confidence of our youth. Indeed, such pervasive adult "protection," coupled with the omnipresent use and overuse of cell phones and social media, are viewed by some as being root causes of the rise in student depression, so-called anxiety attacks, and suicidal thoughts (reported by an astoundingly high 39% of American high-school youth during a recent 12-month period).

Particularly over the past 6–10 years, Bill has repeatedly seen the results of such societal change, not just in the students but in their families, problems that can be painfully acute for first-year students and third-year junior college transfers who, before their entry to UCLA, had never before "been on their own." To cut to the chase, the basic problem is that the students feel isolated, alone, without their previous support system – which, of course, they took for granted. That's a problem that can be handled by trying to build the students' self-esteem and to suggest ways to better use their time. But in the really heart-breaking cases, the problems stem from their off-to-college departure from their home and the toll that takes on their parents, primarily their mothers, who had invested some 17 years in their upbringing and, once their charges are "out of the nest," have lost their own sense of self-worth. Bill has been told of court-battles and drunken family fist-fights. Awful stuff! And time and again, the student feels responsible. They cry (the reason Bill keeps a pack of handkerchief-like paper KimWipes on his office desk) and he tries his very best to convince them that "it is not your fault."

Bill can help to solve many such problems, but not all. A case in point: Some 8–10 years ago, during the summer, he received a telephone call from a young woman in Oakland, California, asking for his help. To protect her privacy and that of her family, we'll refer to her here as "Ms. Kim," who explained that she was from a Korean family and that she had "a big problem." Bill did not recognize her name and asked her whether she had been a student in one of his courses. Her reply: "Oh, no Sir – my girlfriend was in your course and told me to call you."

The very next morning, she and her mother and father arrived at Bill's office. He talked first with "Ms. Kim," who explained the problem. An all-star ("straight-A") high-school student two years earlier, she came to UCLA burdened by her parents' expectation that she would "beat the world." She had done well in her first set of courses at UCLA but as she ventured from her dormitory to take her first final exam, she had an anxiety attack. She turned around and went back to her dorm room. She did this again for her next three exams and, naturally, received a grade of "F" in all four courses. She then repeated this syndrome for the following two academic terms, after the second having been placed on "probation" by the university, and after the third officially "expelled." Her parents knew nothing of this.

The following year she returned and, using the tuition funds her parents had provided, she enrolled at UCLA classes via the Extension Program, costs of which are about twice that for a standard UCLA student. She exhausted her funds and slept on the floor at her girlfriends' apartment, eating whatever was left over from their nightly dinner. But even as an Extension Student, she repeated the same scenario – attend the classes, do the reading and the homework, take the mid-term exam, do excellently, but fail to show up for the course-ending final exam. As before, she continued to fail all of her courses. Two days before she called Bill, her grade-report had come to her home. But this time, unlike earlier occurrences, it was opened by her mother who, seeing her grades and her enrollment via the Extension Program rather than the university, "went ballistic."

OK. "Ms. Kim" had now educated Bill about this horrendous situation. He then invited her mother into his office to hear her view of the situation. She immediately began to cry – the few words she uttered were in Korean, and he does not know whether she could speak English. She was enormously distraught and went through a third of his box of KimWipes. Tough stuff! Bill then invited her husband into his office. He was, to put it mildly, irate. He spoke English, and Bill soon discovered that "Kim's" father knew many more swear-words than did he and that he put them together in novel combinations, literally yelling at Bill as a "f_ _ _-nosed hole" who had allowed his daughter to fail in this "filled-s_ _ _ hell-place!"

Wow! Before that day Bill had never even met this student yet he, a mere UCLA professor, was now pegged as being the root of her problems. What was he to do? He took his three visitors into the adjoining CSEOL library as he thought through their situation, which was of overwhelming importance to all three.

Bill truly felt for these folks – this was an all-too-human, heart-rending situation – but he concluded that it was way beyond his ken. It seemed obvious that "Ms. Kim" had some sort of mental problem that he could not solve. He therefore called the UCLA student psychiatric service and talked with psychiatrist who very kindly told him: "Bill, don't be concerned. I see three or four such students each academic quarter. Bring 'em over and I'll talk with them." He did – and the doctor did. (Two years later, "Ms. Kim" came to visit Bill, now graduating on time thanks to the professors who permitted her to remove her "F" in their classes by taking a make-up final exam – and, most importantly, thanks to the help provided by the kind doctor.)

What Bill learned from this episode is that he, by himself, cannot do it all, and that it is important for him to realize the limitations of his knowledge, experience, and abilities and to then have the good sense to turn to relevant experts to bail him out.

Unfortunately, however, such situations have continued at an accelerating pace. This past year, for example, two young women in his General Education Course of 120 students had similar problems – one an Aeronautical Engineering student who visited him 12 times during the 10-week academic quarter (whom he managed to "save" by restoring her self-confidence); and the other, the daughter of Russian immigrants and a student with whom Bill was able to converse in her native language (much to her delight), who, because of her anxiety attacks, has taken a six-month "medical leave" from UCLA.

"Ms. Kim's" conundrum and that of the budding Aeronautical Engineer came to a happy ending as have many of Bill's less intense interactions with students. But being a teacher and participant in teacher–student interactions is not nearly as straightforward and simple as outsiders might imagine. Be that as it may, it is clear to Bill that over the years, the students have been exceedingly kind to him, as is reflected in their evaluations of his teaching – a long-term record of performance regarded by former provost Brian Copenhaver to be "unparalleled" at UCLA, an assessment more recently reiterated by senior dean of the College of Letters and Science, Joseph Rudnick.

Bill has also been fortunate to have mentored a great group of wonderfully gifted, highly talented graduate students over the years. Three were former Peace Corp volunteers (two stationed in Nepal, one in Brazil); several have gone on to have successful careers in major petroleum companies (Chevron Oil Field Research Company and Conoco Inc.) or have served as research scientists at CSIRO and the Bureau of Mineral Resources, Baas Becking Geobiological Laboratory, Australia. Two have held positions in aerospace enterprises, at NASA's Manned Spacecraft Center in Houston and at SpaceX (Exploration Technologies Corporation, Hawthorne, California). One serves as water resources manager and chief hydrogeologist of the Golden State Water Company; another is vice-president and chief financial officer (CFO) of Galitor Trading, Inc. (Santa Monica, CA); yet another has had been an outstanding high-school teacher in Los Alamos, New Mexico; and one has become a multiple award-winning author of thriller/suspense novels (writing stories said often to be compared to those of Michael Crichton and Dan Brown). Many have had distinguished careers in higher education (in the United States at Beloit and Gustavus Adolphus Colleges; at Georgia State, Ohio State, Penn State, San Diego State, Texas

Tech, and Tulane Universities; at the Universities of Cincinnati, Iowa, Notre Dame, and UCLA; and in other countries at the University of São Paulo, Brazil, and the University of Tasmania, Australia). Several have been at the forefront in applying novel analytical techniques to the study of ancient fossils – most notably, transmission electron microscopy (TEM); confocal laser scanning microscopy (CLSM); secondary ion mass spectrometry (SIMS); biochemical paleogenetic ancestral sequence reconstruction (ASR); and Raman and fluorescence spectroscopy – and have emerged as leaders both in Precambrian Paleobiology and Astrobiology.

Bill is proud and pleased about the success of each and every one of them. And, over the many years, all have remained close friends of he and his wife Jane – and, he would like to think, have benefited from their rigorous broad interdisciplinary UCLA education.

7.6 WHY THIS "SIDE-TRACK" ABOUT TEACHING?

For what reason, you may ask, is this brief discussion of Bill Schopf's teaching efforts included here? After all, it seems only marginally relevant, if that, to unraveling the history of Earth's earliest life and how Darwin's quandary has been put to rest.

There are two principal reasons. First, the discussion highlights the fact that worldwide, such science is conducted largely at academic institutions where its practitioners have countless demands on their time that may, but often do not, have relevance to their particular research interests. If one wants to understand the workings of science – and how this paradigm shift and its subsequent "articulation" occurred – this is a salient fact that needs to be taken into account, if for no other reason than to help explain why progress in the science has occurred more slowly than one might prefer. And second, because as you will have now come to realize, Bill is particularly concerned about the future – today's students, tomorrow's leaders who may read this book– and he thinks it could be useful to them to know that teaching is not merely some "time-consuming unwelcome distraction" that inhibits their research. Rather, college and university teaching is an activity that can feed back into a scholar's work by improving productivity and overall excellence while at the same time bolstering self-esteem, both of the teacher and the students, and by helping our society to improve.

In those regards, four major lessons come to mind:

1. Education is a two-way street; the students learn from the teacher, but the teacher also learns from the students – and Bill's students have taught him a lot!
2. Bill's role has been to help the students learn, not just about academics but also about themselves, their future, their path in life. Most have been receptive, and more than a few have followed his advice – a bit of a surprise, perhaps, given that he is old enough to be their grandfather!
3. Undergraduate students, especially 1st- and 2nd-year students, are appreciably more malleable than their older graduate-student compatriots. Indeed, Bill learned early on that the character and work habits of graduate students are typically rather firmly set by the time they enter graduate school, and

that about the best he can hope for as their mentor is to identify one, two, or at most three areas of needed improvement and to work continuously on those issues throughout their graduate careers. Even then, he sometimes misses the mark; as one his early students told him years later: "The best thing you taught me was how to write" – when Bill, of course, had imagined that he had been teaching about how to be a decent person and do excellent science!

4. Finally, it pleases Bill immensely that from time to time, his ideas have led to absolutely first-rate Master's and Ph.D. theses that have moved the field in beneficial directions. Phrased another way, it is abundantly clear that not one of us can "do it all." Excellent students, such as those Bill has been so fortunate to have mentored, have extended and amplified his reach. This is not to be taken for granted. The students deserve their due. Bill is grateful.

8 This Science Over the Years

8.1 PERILS OF GEOLOGICAL FIELD WORK

All major universities highlight the research activities of their faculty and students, pleased by the national and international attention (and increased donors' funding) garnered from favorable media coverage. The University of California Los Angeles (UCLA) is no exception. For many years, the principal media contact for the UCLA Physical and Life Sciences has been Stuart Wolpert, who has been a long-time personal friend of Bill's since they first met when Stuart was a student in the Honors Program for which Bill served as dean. Knowing of Bill's travels and a few of his treks into the "untamed wilds," Stuart – like many who have not themselves participated in such activities – harbored a romanticized view of the adventures, telling him at one point that "Gosh, you're just like a real-life Indiana Jones."

Actually, of course, "real-life" geological field work is not at all like that depicted in the movies. In fact, it's a whole lot more like being a grubby and unwashed Boy Scout – it is lots of fun and lots of hard work too, but it is not at all like being a swashbuckling Harrison Ford! And, it can be dangerous. Below are outlined three episodes during which Bill came close to being seriously injured or perhaps killed.

In the early 1970s, when Bill returned home from one of his early trips to the Soviet Union, he found a letter taped to the front door of his home asking him to contact the producer of NOVA, a nation-wide science program on the PBS television station WGBH Boston. The program wanted to fly him down to Baja, Mexico, to film one of his field projects studying modern mat-building (stromatolite-forming) microorganisms.

Bill agreed and boarded a commercial flight from Los Angeles to San Diego, where he was met by a helicopter pilot and recent veteran of the Viet Nam War, like Bill in his late 20s, who was flying a two-seater single-engine fixed-wing Cessna 150 in which they headed south to Baja. During the uneventful two-hour flight, the pilot regaled Bill with war stories (mostly about the pilot's buddies, women friends, and marijuana), and about half-way through the flight, he invited Bill to take control of the plane. That was plenty exciting. Bill had never before (or since!) done such a thing. He took the "stick" and almost immediately the plane went into a steep "nose-dive" – he'd had no coaching, and the controls were far more sensitive than he had imagined. He quit. The pilot took over and they landed safely in a farmer's field a couple of miles from the field site. The film crew was already on hand, having driven down from San Diego.

They filmed for a couple of hours and finished their work, and then the pilot and Bill reboarded the Cessna and taxied to the far end of the field where the pilot revved the motor. The field was short, not more than half the length of a football field.

They hurtled along, gaining speed, but the wheels then hit a patch of sand. The plane jolted and slowed. It was too late to abort the take-off and the plane was about to hit a set of six power-lines at end of the field. The pilot jerked back on the stick, hard, and up they soared up at a precipitously steep 45° angle. Neither of them said anything for three or four minutes until Bill muttered, "Wow, that was a close call." The pilot's rejoinder: "Ahh, that was nothing – you should have been in Nam." (A few weeks later, a view of the film crew's outtakes showed that they had cleared the power-lines by a scant few inches.)

A second such harrowing episode occurred some years later when Bill was doing field work in the eastern part of central Australia, riding "shotgun" next to the driver, a geologist from the Geological Survey of Northern Australia. It had rained for several preceding days, unusual for the central Australian Simpson Desert, but it had not affected their earlier work. They were speeding along at about 70 miles per hour when they came to the top of a small rise in the road and saw before them a vast sea of water. The water was probably only a couple of feet deep, but to plow into it at that speed would have been catastrophic. The driver turned the wheel sharply to the right; the field vehicle started to skid, to slide; and then tipping sharply toward its right side, with the left tires off the road, the vehicle did a "wheelie," slipping and sliding down the edge of the road for some 75–100 meters before righting itself as they ascended the next small rise. (To say the least, this was a "close call.")

Yet a third episode – to Bill particularly terrifying – occurred when a team of geologists he was co-leading, six or eight in all, were doing field work in the Hamersley Iron Basin of Western Australia. This is beautiful country, with bright cherry-red 2.5-billion-year-old rocks and deep river-carved gorges. The rock strata flanking the gorges are ledged, and in places, the canyon walls are easy to climb due to the differential weathering of their alternating iron-rich and iron-poor layers seasonally deposited billions of years ago (Figure 8.1). The team had trundled down a dry river bed and began, one behind another, to climb out. As a leader of the group, Bill went last, slowed only by the 60–80 pounds of rocks in his back-pack.

Eventually, Bill reached the top – or, more precisely, about eight feet from the flat land at the top of the gorge. The others had gone on before him, but all was well until he made the stupendously stupid mistake of looking down to the bottom of the gorge. There beneath him he saw a vertical drop of 800–1,000 feet. His knees and legs instantly turned to jelly. They lacked all feeling and literally had no strength. Bill had never before experienced such a thing. Oddly, however, he wasn't all that scared. Rather, he reasoned through the situation and thought to himself this is just a simple problem of "'mind over matter' – your mind tells your legs what to do, so, for Gosh Sake's, get 'em working!" He rested and tried that solution: "Mind, get your legs to work!" No response. He waited and tried it again. Nothing happened – his legs were numb, completely non-functional, and his knees were buckling. He then called out, as loudly as he could, for the others in the team to come to his rescue – but they were long gone and did not hear his plaintive cries. He was on his own.

Now fully aware of his dire predicament, he thought through the problem and bawled himself out: "Solve this, you stupe, it's your own darn fault!" The ledges above him were separated by three to six inches. So he elbowed his way up, first with

FIGURE 8.1 Vertical canyon in 2.5-billion-year-old iron-rich rocks of Western Australia.

the left elbow and then the right, onto the next ledge, dragging behind his dysfunctional legs. He did this again, and again, and again, ledge after ledge, and slowly but surely elbowed his way up the side of the gorge (much as he done as a 5th-grader when he shimmied up the sides of a toilet stall to unlatch a window so that the next morning he could let his pals in for basketball practice). After about 15 minutes, he made it to the top. He lay there on his back, panting, and again bawled himself out – he'd been dumb; he should not have looked down! But he also thought: "The damage is done – it's behind you now – move on." Miraculously, his legs had now recovered and his knees were now just fine. He raced ahead, caught up with the others and, embarrassed by his stupidity, did not tell them much about his horrendous plight. *Yes, Oh Yes ... Bill has lived a lucky life!*

8.2 THE WONDERS/TERRORS OF GEOLOGICAL FIELD WORK

Such "death-defying" adventures are not uncommon to geologists. Field work can be perilous, as most geologists know all too well. But such events also carry object lessons from which all can benefit.

The fact is that each of us humans, we so-called "intelligent" bipedal primates, is vulnerable, sometimes because of our own errors and other times due to events over which we have no control. But it does no good to fret about such perils. If one can, the goal is to correct the errors. But if a mishap it is not one's fault, then "thank your lucky stars" that it turned out OK. Moreover, and despite the "me-first gene" embedded in each of us social primates, none of us – not even the admired, most successful among us – actually matters all that much. Of course, each of us matters to our family, friends, and colleagues – that too is embedded in our genes – and what we do during our lifetimes can matter greatly in its particular time and social context. But in the great sweep of human history, for each of us, our fleeting time on planet Earth is unlikely to have a huge and lasting impact. And time after time, science proves the point. If there is a major unknown that has been recognized but remained unsolved – such as the history of life in deep time – someone at some time would ultimately come along and present its answer to the benefit of us all. That is the way of science, the way that generation after generation learns more and more about the reality around us.

There is one last take-home lesson from such adventures, namely that the gloriously romantic myth of geological, paleontological, anthropological, or biological field work is just that – a myth – romanticized by others who have no personal understanding of what it actually entails.

8.3 THE "UPS" AND "DOWNS" OF SCIENCE

As you might well imagine (especially if you know something about the workings of science), Bill Schopf's contributions have not always been greeted with immediate applause. Indeed, in 2002, after nearly a decade of widespread acceptance, one such finding – his 1993 report of discovery of especially ancient, 3,465-million-year-old microbial fossils from the Apex chert of Western Australia – engendered a good deal of divisive controversy. In 2006, that controversy, coupled with Bill's 1996 skeptical evaluation of NASA's hopeful report of the discovery of "possible microfossils" in a meteorite from Mars, even led him to be depicted as an "intimidating devil's advocate" in a book-length discussion of these studies (written by an author who never personally interviewed him).

Until now, Bill's public reaction to such matters has been stoic silence. Why? The answer is simple. He has the notion that in science, *firm facts will always win!* In other words, he really *does* prefer "to take the high road," to let the facts speak for themselves and to not engage in inappropriate (and in his view self-defeating) *ad hominem* attacks on others. And unlike some workers, he truly does not believe that serious scientific questions are properly decided in the too-often emotionally charged "Court of Public Opinion." To him, it is the facts of the science that matter – not he, nor the naysayers, nor whatever preconceived notions one might wish the facts to support.

But Bill has learned that his "above the fray approach" is sometimes rankling to others. A case in point: The December 2011 issue of *Discover* magazine, in a short blurb about the Apex microfossil controversy in which the author of the piece describes Bill Schopf as "one of paleobiology's great lions," his prime critic (Professor Martin Brasier of Oxford University) is quoted as suggesting that arguing against Schopf "was like tackling Jesus or Moses" – the "tackling" presumably referring to English "football," known in America as soccer. Whatever Brasier's motivations may have been – there is no way to know – he certainly placed Bill in sterling company (even though upon further reflection, Bill jokingly thought to himself that Brasier's list might also have included Buddha, Confucius, and Muhammad!)

The following narrative is a synopsis of the history of this decade of acrimony. The narrative deals first with the 1993 Apex paper and the long controversy over these especially ancient fossils and then moves on to outline the mid-1990s NASA "Martian possible fossils" dispute, which, because of Bill's skepticism of the claim, placed him in the "Bull's-Eye" as an object of resentment and derision. These two stories are then brought together with a short discussion of the 2006 book that highlights the controversies. Finally, the narrative addresses the latest (2018) findings which show how and why the dispute has been resolved, decided at long last on the basis of firm facts.

Others knowledgeable about these events might well have recollections that differ from those here recounted. Such differences regarding events of years ago would be understandable, are likely to be minor, and should not be disparaged. Yet despite any such differences, the final conclusion would remain unaffected – in this and in any other controversies in science, *firm facts will ultimately win*. They must! There is only one reality, and the task of the practitioners of science is to describe and understand that reality as thoroughly and dispassionately as they can.

8.4 A GREAT CONTROVERSY IN THE SCIENCE: THE EARLIEST RECORDS OF LIFE

In 1993, Bill Schopf wrote a paper published in *Science* titled "Microfossils of the Early Archean Apex chert: new evidence of the antiquity of life," which presented analyses of 175 microscopic fossils and their component 1,877 cells in the 3,465-million-year-old Apex chert of Western Australia – the oldest diverse assemblage of fossils then and now known from the geological record (Figures 8.2 through 8.4). Geologically initially mapped as having been deposited in a shallow marine setting, the fossiliferous locality of the Apex chert has more recently been reinterpreted to be a fractured hydrothermal vein deposit. Although at first glance, such a hydrothermal environment may seem an unlikely setting for the preservation of delicate fossil microbes, it is worth noting that microorganisms similar in shape and size to the Apex filaments are actually rather common today in hydrothermal settings, and that cellular filamentous microbes similar to the most abundant of the described Apex taxa (*Primaevifilum amoenum*) have long been known to occur at deep-sea thermal vents. In addition, other workers have discovered additional chert-permineralized fossil filaments, including specimens so similar to those of the Apex deposit that they were assigned to two of the Apex taxa, in three additional hydrothermal units almost as old as the Apex chert that occur in the same region of northwestern Western Australia. Moreover, and perhaps most importantly, the Apex fossils occur in rounded carbonaceous clasts, bits and pieces of organic-rich material that may have been emplaced in the Apex chert from an earlier setting decidedly different from that suggested by the rocks in which they now occur.

The 1993 paper was well received. The discovery there reported, included also in Bill's 1999 freshman-sophomore-educated-lay-person (and national prize winning) volume *Cradle of Life*, was widely accepted and featured in textbooks and museums of natural history across the globe. It was therefore quite a surprise when eight

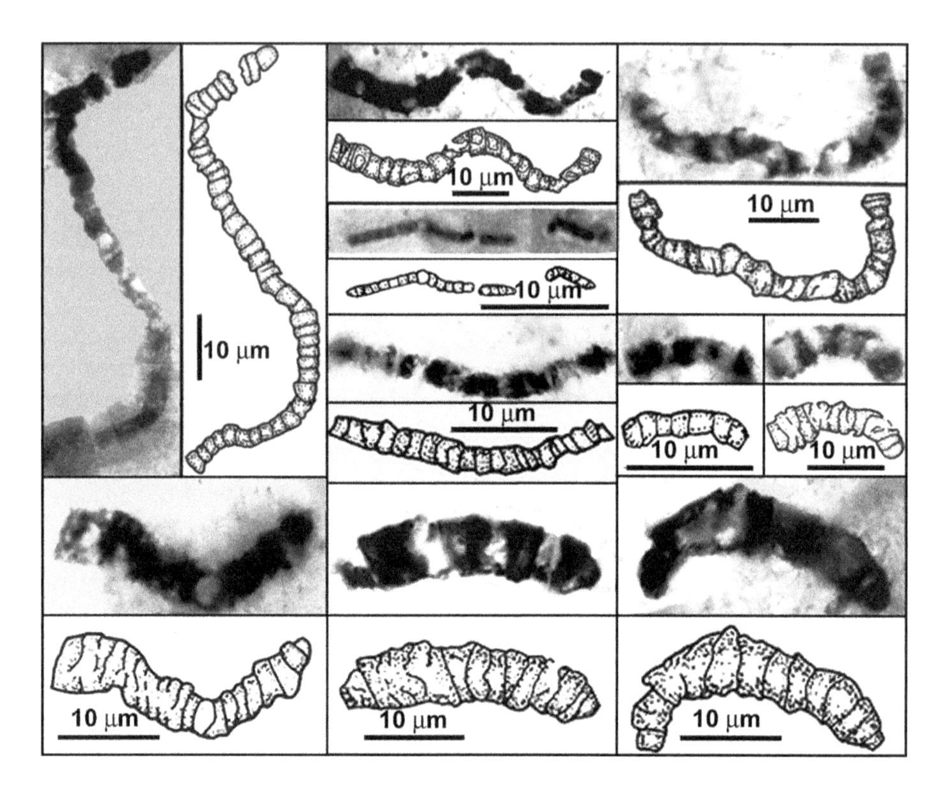

FIGURE 8.2 Optical photomicrographs and drawings of representative microbial fossils of the 3,465-million-year-old Apex chert of Western Australia.

years later – near the beginning of 2001 – Bill received an unpublished manuscript from the journal *Nature* that questioned his 1993 interpretations. In particular, the manuscript asserted that the Apex fossils were not, as Bill had inferred, composed of coaly kerogenous organic matter but were instead chain-like aggregates of such "opaque minerals" as pyrite [FeS_2], chromite [$(Fe, Mg)Cr_2O_4$], ilmenite [$FeTiO_3$], and various iron oxide minerals that the authors claimed to have "identified by optical microscopy." This seemed odd – such minerals are by definition opaque, entirely black in transmitted light, and because they do not transmit plane-polarized light, they are notoriously difficult to definitively "identify" by optical microscopy unless they exhibit sharply defined euhedral crystal faces, which the mineral grains of the Apex chert do not. Nevertheless, Bill was more than a little concerned. Had he been mistaken?

Bill was prohibited by *Nature* from providing a review of the unpublished manuscript – which, because of its presentation, was regarded by journal-editor Henry Gee to be a "personal attack." Fortunately, however, two years earlier, Bill had devised a method (Raman spectroscopy) to analyze, *in situ*, the chemical composition of just such rock-entombed tiny fossils (a new advance, not available in 1993 when the fossil assemblage was first described). So, soon after his receipt of this manuscript, Bill obtained the specimens he had previously described and archived at London's

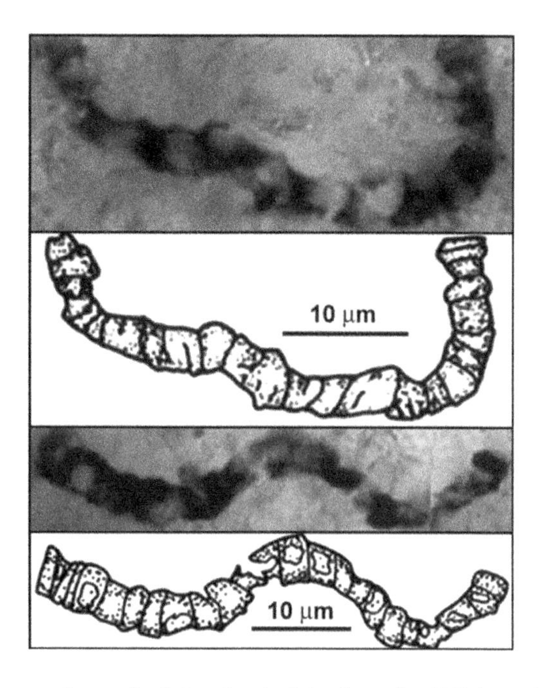

FIGURE 8.3 Two specimens (and drawings) of the Apex fossil *Primaevifilum amoenum*.

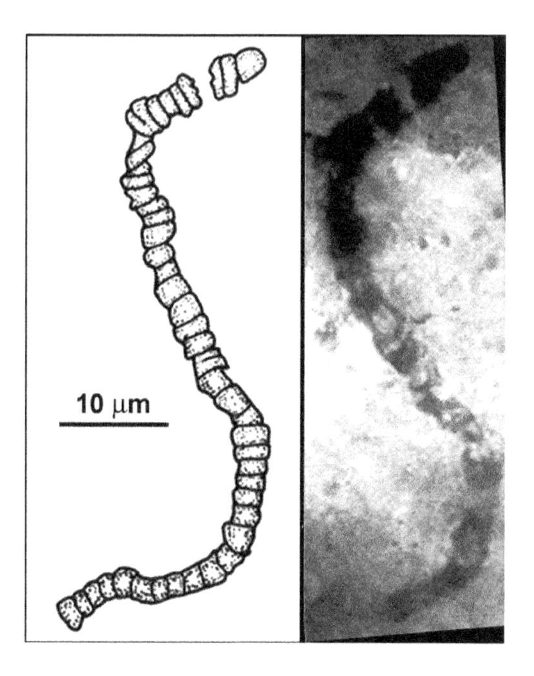

FIGURE 8.4 A long, many-celled specimen of the Apex fossil *Primaevifilum amoenum*.

Natural History Museum and then used Raman to establish, much to his relief, that of the specimens in the Apex assemblage, the many that he had originally inferred (on the bases of color, texture, and cellular and organismal form) to be carbonaceous were, by direct *in situ* analysis, demonstrably composed of carbonaceous organic matter, not opaque minerals. "Ha," he thought, "this problem will now end, facts will win!" (As it has turned out the facts *have* ultimately "won" – but it has taken more than a decade.)

Bill did not then – nor does he now – know the motive(s) of the critics. Honest skepticism? Lack of knowledge and experience? Hoped-for personal fame? Or perhaps the root lies in the first of writer and futurist Sir Arthur C. Clark's triad of sequential responses to game-changing ideas in science: (1) "I never heard that before, it must be wrong," followed by (2) "It might be right, but it's probably not important" and (3) "It's obvious – I knew it was a good idea all along!"

In June 2001, Bill and his colleagues prepared a manuscript reporting their new Raman-based findings, submitted it to *Nature* and, on that same day, sent it to Oxford University professor Martin Brasier, the lead author of the unpublished manuscript Bill had previously received, so that he and the Oxford team would be aware of the new findings. Brasier did not respond. Bill waited three months. Then, thinking that for some unknown reason Brasier had not received the earlier missive, Bill again sent a copy by FedEx, for which the recipient is required to sign. Several weeks later, Professor Brasier replied, indicating that he would "respond in due course." He never did.

Meanwhile, *Nature* editor Henry Gee, who wanted the Schopf and Brasier papers to be published "back-to-back" (presumably to promote the prominence of his journal) then held the already accepted Schopf et al. paper for eight months to allow the Oxford group sufficient time to check the results reported. They did and found the Schopf analyses to be correct – the fossils *are* composed of coaly organic matter (just as are virtually all other cellularly preserved authentic Precambrian microscopic fossils).

Confronted with this fact, the Oxford team then changed its story, asserting that the fossils are composed of "abiotic graphite" – that is, composed of the crystalline, "pressure-cooked" carbon mineral graphite that had been produced by entirely *non*-biological processes. They are not. As the Raman data establish, both in Schopf's analyses and those of the Oxford team, the fossils are in fact composed of a biologically produced carbonaceous coal-like material known as *kerogen* (its biogenicity being further confirmed in 2009 by in-depth chemical analyses by Barry DeGregario and his colleagues at Arizona State University, the State University of New York Plattsburgh, and Brookhaven National Laboratory).

But to address the discrepancy between the Raman-based identification of kerogen and the abiotically formed "graphite" interpretation preferred by Brasier and his co-workers, their paper, finally published back-to-back with the Schopf et al. article in 2002, referred to this carbonaceous matter as "amorphous graphite," ignoring the crystallinity required for the term *graphite* to be appropriately applied. In follow-up papers, they clarified this revision, creatively indicating that they were using this term "*sensu lato*." In other words, they used the term *graphite* according to their own newly coined broad definition – rather than in its universally accepted sense (*sensu stricto*) – a recasting of the accepted definition of this well-known and rather commonly occurring crystalline carbon mineral that would better suit their needs.

The Oxford team also, both then and repeatedly later, persistently asserted that Bill Schopf's 1993 paper identified the fossils as "cyanobacteria" – an evolutionarily relatively advanced group of oxygen-producing photosynthetic microbes – which the paper certainly did not. Instead, because Bill thought that the information available was insufficient to be certain of the biological affinities of the incompletely preserved Apex specimens, the 1993 paper devoted three long paragraphs to justifying his decision to identify them merely as "*Bacteria Incertae Sedis*," bacterial microbes of undetermined and unspecified biological relations. And, as Bill had taken pains to point out, he particularly did *not* want to claim that they were cyanobacteria – the dominant members of a great many younger Precambrian biotas – because such an assertion would have pegged these especially ancient fossils as oxygen-producing photosynthesizers and would thereby imply that early evolution had proceeded farther and faster than was actually known. In short, Bill's carefully thought-out goal was to not mislead the scientific community – it being quite sufficient, he thought, to simply establish the existence of this large assemblage of morphologically diverse fossil microorganisms so early in the history of the planet. For reasons not easily fathomed, the Oxford team ignored this discussion.

During that same period Professor Karl Stetter of the University of Regensburg, Germany told Bill that Brasier had described him at a scientific meeting in England as a "charlatan" and the perpetrator "of the greatest hoax in the history of science" – comments that led another of Bill's friends to suggest that that Brasier had perhaps been influenced by the "typical hyperbole of the famous Thursday evening Oxford Union debate program." And in 2005, in Bill's presence at symposium at the Royal Society of London, Professor Brasier publicly proclaimed that "all evidence of life older than two billion years is hearsay!" – an assertion that world-famous Harvard geochemist Heinrich D. Holland immediately questioned because he knew that it denigrated some 40 years of accumulated evidence (fossils, stromatolites, carbon isotopes, organic biomarkers, and so forth).

Bill remembers thinking at the time that Brasier's unsupportable assertions smacked of what his former department colleague George Wetherill had dubbed "the young man's disease" – an attempt to win an argument by obfuscation and misrepresentation rather than by evaluation of the available facts. But Brasier was not a "young man," and in this instance, the situation was compounded by his evidently limited experience in the relevant science. This inexperience was not limited to Raman spectroscopy (of which Brasier seemed to have had no knowledge before his receipt of the Schopf et al. unpublished manuscript) but extended to Precambrian microscopic fossils generally – which he had apparently never previously investigated – and to the biology of comparable modern microbial assemblages (which, similarly, he had evidently not previously studied). And his inexperience was further amplified by his misapplication to the Apex fossils of a recently developed (and new to him) auto-montage confocal microscope, which artificially flattened three-dimensional images into a single plane (causing him to misinterpret superimposed individual minute unbranched Apex filaments as being interconnected and thereby composing an implausibly evolutionarily advanced single multi-branched specimen).

A few years later, Brasier visited McGill University in Montreal to confer with Precambrian Paleobiologist Hans Hoffman, the world's leading expert on the series of changes that occur during the preservation and "geological aging" of fossilized microbes. This detailed knowledge had led him to divide such fossil-like objects into three categories: (1) *bona fide* fossil microbes; (2) *dubiofossils*, microscopic "fossil look-alikes" of possible but unproven biological origin; and (3) *pseudofossils*, non-biologically formed objects that could not be regarded as fossils. Having personally studied Bill's Apex fossils, Hans regarded them to be assuredly *bona fide.* As Hans later explained, he made a serious well-meant effort to teach Dr. Brasier about the degradation sequence (technically, the "taphonomy") of fossilized microbes, but Brasier would have none of it, evidently having already made up his mind. As Hans later phrased it, his explanation had fallen on "deaf ears."

Even later in this ongoing saga, this time again in Bill's presence, Brasier publicly touted his expertise by indicating that he was a "card-carrying micropaleontologist" – which in fact he was, and author of a micropaleontology textbook as well. But his statement was misleading, implying to the uninitiated that he was expert in the study of "small fossils" – presumably like those of the Precambrian – whereas in fact it actually meant that he, like virtually all other micropaleontologists, was experienced in the identification and study of the shelled microscopic fossils of very much younger geological time (foraminifera, calcareous nannofossils, ostracodes, and the like, which did not exist in the Precambrian).

Professor Brasier's co-workers in this venture were evidently as personally committed to this effort to discredit the ancient Apex fossils as was he. Shortly after Bill and his colleagues had used their newly introduced Raman technique to document the molecular chemical composition of the rock-entombed Apex fossils and during the period when their *Nature* paper had been accepted but was still unpublished, Bill attended a meeting of a NASA study-group plotting a course for future investigations of current or past life on Mars. Among the 12–15 attendees was recent Ph.D. recipient Andrew Steele, an affable and charismatic if outspokenly opinionated recent appointee to the staff of the Carnegie Institution for Science in Washington, D.C. (at that time, "the Carnegie Institution of Washington"). Bill gave a talk outlining his group's new Raman-based results on the Apex assemblage and later, at Dr. Steele's request, privately discussed the results with him in more detail.

Unbeknownst to Bill, Andrew Steele (known to his friends as "Steelie") – a transplant to the United States from Portsmouth, England – was a co-author of the forthcoming Brasier et al. 2002 paper. He thus had a vested interest in pumping Bill for details of his work. But Dr. Steele did not divulge his participation in the competing project – as best Bill could tell his inquisitor just seemed "interested" in the new findings. (HA! Evidently Bill "got taken." In retrospect, his error was in assuming – as he has done far too often – that others in science play by the same rules as does he.)

Later, following additional such "anti-Schopf" incidents, one of Bill's acquaintances commented: "The vitriol of these naysayers is palpable." And so it seemed – Professor Brasier and his colleagues had an "axe to grind." But Bill Schopf has never understood exactly why. In fact, Bill never personally "knew" Martin Brasier (who died in 2014 in an automobile accident) or the other principals involved in the work of the Oxford team. Yes, Bill spoke once with Professor Brasier and was present, two or three times, at symposia in which Brasier also participated. But as best Bill can recall, he only ever said six words to Brasier when, at a 2005 meeting of the Royal Society, Bill walked over and introduced himself: "Hello. My name is Bill Schopf." (Brasier's vehement and seemingly disdainful five-word rejoinder: "I *know* who you are!")

Interestingly (and perhaps of some note to philosophers or historians studying the workings of science), skepticism about the biological origin of the Apex fossils continued – devoid of its earlier *ad hominem* attacks – carried forward by others. Given the opening salvo provided by the 2002 Brasier et al. *Nature* paper, in subsequent years various other naysayers entered the fray.

A few months later and again in 2003, one group published papers suggesting that the Apex fossils might be composed of barium carbonate crystallites like those they had synthesized in laboratory experiments. (Won't work ... such crystallites are mineral, not carbonaceous, and they are solid, not composed of organic-walled cells; and, like the "graphite pseudofossils" postulated by the Brasier group – but unlike the literally millions of authentic microscopic fossils known from Precambrian sediments – "barium carbonate pseudofossils" have never been found in the geological record.)

Then, a couple of years later, in his pre-publication review of one of Bill's papers, a US Geological Survey geochemist opined that the manuscript should be rejected because "As everyone knows, there are lots of non-biologically produced organic pseudofossils in Precambrian rocks." Bill wrote him and thanked him for his observation, but asked him to please provide some references to such studies because Bill had no knowledge of them and wanted to learn in case he later wanted to write a paper addressing the problem of distinguishing the bogus from the *bona fide*. The USGS geochemist, like Brasier, never replied.

In 2009, there was another suggestion, this one implying that the Apex fossils are needles of silica (i.e., quartz) or hallosite, a well-known clay mineral. (Again, this won't work ... such needles are mineralic, not carbonaceous; they are solid, not cellular; and they are some 40 times smaller than the reported fossils. Evidently realizing this, these authors backed off, writing the disclaimer that "These observations are not applicable to the [Apex] microfossils – because we did not observe carbonaceous filaments.")

In 2011, there was yet a fourth proposal, this one based on Raman spectroscopy and again suggesting that the fossils are mineralic, in this instance quartz-hematite vein-fillings. This episode was to Bill particularly dismaying, not only because by this time he had published an article specifically warning about such misinterpretations, but also because it seemed evident that the lead author of this paper, Dr. Craig Marshall – an Australian who as a post-doc had worked with Bill's long-time friend Malcolm Walter at the University of New South Wales in Sydney – had been appointed to his current position at the University of Kansas at least in part because of the work pioneered in Bill's lab.

When one of the Kansas faculty members informed Bill that they had hired Dr. Marshall to join their department Bill wrote a warm letter to Marshall, welcoming him to the United States and telling him that he "would be glad to help him in any way [Bill] could." Marshall never responded either to this or two later missives.

The lack of replies remains a mystery, although one of Marshall's students later told Bill that "Oh yes, he received those letters. He told us about them."

In 2005, Dr. Marshall was a pre-publication reviewer of what turned out to be Bill Schopf's landmark paper documenting the paleobiological usefulness of Raman to analyze the chemistry of microscopic fossils embedded in ancient rocks – a technique now used worldwide. In his review of this manuscript, Dr. Marshall opined that this new application of Raman was "absolutely impossible" because in his view, "Raman is a surface-only technique that cannot see inside rocks." At that time, Marshall was relatively inexperienced in the study of ancient rock-entombed microscopic fossils – and he was way off-base. Since then, he has used the technique on numerous fossils – with admirable success.

As for Dr. Marshall's 2011 criticism of the 3,465-million-year-old Apex fossils, he was simply mistaken. Instead of using the Raman spectral "G Band" of kerogen (at 1,600 cm^{-1}) as Schopf and Kudryatsev did in their work, Marshall's study centered on the Raman "D Band" of kerogen (at 1,350 cm^{-1}), which is fairly close to a broad band of the iron-oxide mineral hematite (at 1,320 cm^{-1}), and for this reason suggested that fossils are mineralic hematite vein-fillings. Moreover, the objects Marshall and his co-authors reported are mineral, not carbonaceous; are planar sheet-like fracture-fillings rather than cylindrical garden hose-like microbial filaments; are straight, not sinuous; are solid, not cellular; and are much longer and three to eight times broader than any of the Apex fossils – quartz-hematite vein-filling pseudofossils that markedly differ from *bona fide* Precambrian filamentous microorganisms.

Most troubling, one might suppose, is that none of the authors of these various follow-on alternative hypotheses seems ever to have examined and analyzed the originally described Apex specimens, which, in 1993, Bill Schopf archived at London's Natural History Museum so that they would available to all serious workers (Figure 8.5). Instead, they have preferred to collect their own samples of the Apex chert from the "Schopf locality" – a remote, hilly, large football field–sized area now protected in the Chinaman Creek Geoheritage Reserve, created in 2009 by the government of Western Australia to manage access to the site. As anyone who has ventured to the site would be aware, it is geologically complicated by the presence of other strata, by boulders eroded from associated units, and by interlaced intruding chert dykes unrelated to the Apex unit.

Moreover, recollection of samples from the exact site, the same fossil-bearing rocks collected by the Precambrian Paleobiology Research Group (PPRG) in 1982, has been made even more difficult by the deficiencies of the technology then available. At present, by use of global positioning satellite (GPS) systems, field geologists can pinpoint a specific collecting locality to within a radius of 10–30 meters (and by use of multiple satellites to triangulate the site to within perhaps 3–5 meters). But in the 1980s, GPS technology did not exist. The best one could then do was to use the highest resolution topographic map available, identify the locale relative to the local geographic features, and record its position by a pin-prick in the map. Though at the time this was the "gold standard" in such work, it was far less precise that one would have preferred. Indeed, when Bill returned to the Apex site 17 years later, in 1999, to collect additional samples from the fossil-bearing chert, it took him nearly three hours to find the precise site and the precise rock – and he succeeded then only

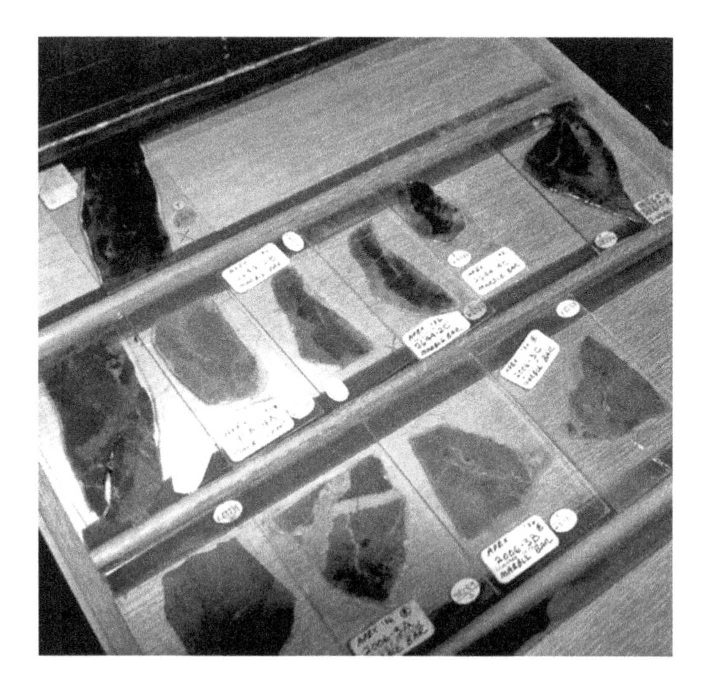

FIGURE 8.5 Fossiliferous thin sections of Apex chert archived at London's Natural History Museum.

because he had brought along 1982 field photos, both of the overall setting and of the specific rock originally sampled. Had Bill been asked to help any of these follow-up workers in their search for the precise fossil-bearing site, he would have been pleased to provide assistance.

This lack of consultation regarding the fossiliferous site and the apparent absence of appropriate attention to the archived specimens is regrettable. Although a hallmark of science is its capability to be self-correcting, meaningful corrections can only come from sound understanding of the bases of the original interpretations.

By now, January 2018, all of the suggestions made by these workers have been shown not to be relevant to the Apex fossils, and all the skeptics appear to have gone silent about the situation. In such instances, the absence of scientist-to-scientist personal communication is regrettable – a lack of colleague-to-colleague private exchange of ideas that Bill tried several times to remedy with his repeatedly unanswered letters. Heavens, all scientists are in the same game, each trying to describe and understand reality, and none is infallible. It would be far better for all involved to be a PPRG-like "band of brothers" helping one other to advance the science.

Nevertheless, as things turned out – as Bill Schopf's 1993 findings were assailed during the first decade of this century – the misguided interpretations of these skeptics raised doubts in the scientific community that then diffused to members of the media who, being unschooled in such studies, quite understandably did not know whom to believe and why.

8.5 NASA'S "AUGUST 7TH EVENT" – "POSSIBLE MICROFOSSILS" FROM MARS

Let us now step back to the mid-1990s, to another episode that got Bill Schopf "in Dutch" with various folks.

In January 1995, as he recalls, Bill received a phone call from Michael Myers, then head of NASA's Exobiology Program in Washington, D.C., asking Bill to fly down to the NASA Johnson Space Center (JSC) in Houston, Texas, to assess some "interesting new findings" (which from Meyers' tone and Schopf's areas of expertise Bill presumed to mean "possible microscopic fossils") in a meteorite thought to have come from Mars. Soon thereafter, Bill received a call from one of the JSC scientists involved in the work that provided further hints about these "unusual structures." As requested, Bill flew to Houston. To his surprise, but great pleasure, he learned there that the Idaho-potato-sized meteorite (ALH84001; Figure 8.6) had been collected a decade earlier, in 1984, at an ice field in the Allan Hills of Antarctica by Roberta ("Robbie") Score (Figure 8.7), who, when Bill met her at JSC, proudly told him that

FIGURE 8.6 Mars meteorite Alan Hills 84001.

FIGURE 8.7 Former UCLA Geology student Roberta Score, collector of ALH84001.

she was a was a graduate of his UCLA department and that, having taken one of his courses, she was "one of [his] students." The two JSC scientists involved in the study, David McKay and Everett Gibson, showed Bill their findings, small (20- to 30-μm-diameter) discoidal objects that he took to be carbonate vein-fillings in a volcanic rock, none of which evidenced obvious evidence of a biological origin. Bill's impression was that McKay and Gibson might have imagined these to resemble flattened poorly preserved remains of shelly Martian protozoans similar those on Earth. Bill suggested that such an interpretation was unlikely and that to prove their case – and to have their results accepted for publication – they would have to produce evidence of biological cellularity and document the presence of organic matter in the structures (i.e., as part of the structures), a one-to-one direct association of cellularity and composition that Bill thought would be key.

To satisfy Bill's request for confirming organic geochemical data, McKay and Gibson presented overlays of the optical photos of the objects showing a close spatial correspondence of the discs to overlying patchily distributed splotches of organic matter (polycyclic aromatic hydrocarbons, "PAHs"), analyses done by Dick Zare at Stanford. At first this seemed convincing. But Bill then asked the telling follow-up question: "Are there PAHs elsewhere, not associated with the discs?" After an exchanged glance the JSC scientists then produced an overlay of Zare's complete data-set that showed the possible biologically (and/or abiotically) derived PAHs to be randomly distributed, not primarily associated with the mineralic discs.

The initial impression of a one-to-one correspondence of the discoidal structures and organic chemistry immediately evaporated as did the possible biological significance of the PAHs (which, because of their geochemical stability are not only a dominant component of fossilized coaly organic matter on Earth but also of *non*-biologically produced organics both in organic-rich meteorites, "carbonaceous chondrites," and interstellar space). Dave McKay is said to have later characterized Bill's assessment during this 1995 visit as "a setback."

The involvement of McKay and Gibson in studies of the Mars meteorite was common knowledge among their JSC colleagues, and Bill's visit had evidently aroused some curiosity – "Why has that Precambrian life guy come to talk with them about their work?" Perhaps foremost among those, Bill learned later, was Robbie Score, the collector of the rock in Antarctica, who knew about his studies from her days at UCLA. In any case, McKay and Gibson pledged Bill to not divulge the reason for his visit or the nature of their findings – an agreement he kept until 1999 when he first wrote about this episode in *Cradle of Life*, four years after his visit to JSC and three years after the NASA press conference and the publication of their paper in *Science*, in which they publicly revealed their subsequent (by then altered) interpretations.

A few months later, Bill and Dick Zare had a good chat at Wesleyan University in Connecticut, where both were lecturing on the same program and Bill explained to Zare the problems about the potential origins of the PAHs he had detected. Zare is an excellent scientist and, as Bill recalls, Zare then indicated that, had he known before what Bill had explained, he might not have agreed to co-author the *Science* paper that ultimately appeared (at that time, to best of Bill's recollection, Zare being a member of the US National Science Board, for whom an unblemished reputation of sound scientific judgment is paramount). Following that, to Zare's credit (whether for this or some other easily conceivable reason), he declined the opportunity to co-author additional works with the JSC group.

During his visit to JSC, Bill suggested to McKay and Gibson that identification of possible fossils is a biological/paleontological problem; that NASA's expertise in such matters was centered at NASA's Ames Research Center in Palo Alto, California; and that they might therefore consider contacting their colleagues at Ames for an appraisal of their finds. Neither McKay nor Gibson evinced any interest in this suggestion, leading Bill to surmise that they may have been concerned that were they to bring their Ames colleagues into the mix, NASA Ames rather than NASA-JSC would reap whatever credit might accrue from their discovery (an internecine competitive mind-set that, at the time, was all too prevalent among the numerous NASA centers).

Bill had done his task and thought that this was all behind him. But a year or so later, he received second call from NASA headquarters' Mike Meyers asking Bill to appear at a planned news conference where he was to publicly evaluate "on behalf the world scientific community" a scientific paper authored by these same JSC scientists and their colleagues, soon to be published in *Science*, reporting discovery of "possible microfossils" in Mars meteorite ALH84001. Bill wanted no part of this. He demurred, telling Mike (whom he had recommended some years before for appointment at NASA HQ) that "extraordinary claims [as that for past life on Mars would certainly be] required extraordinary evidence [which, to the best of Bill's knowledge, the JSC scientists did not have]" – a quote Bill borrowed (i.e., stole) from Carl Sagan, a friend since his graduate-school days. Bill suggested three alternatives to serve in his stead, any of whom, he thought, would be likely to give a more "NASA-friendly" positive appraisal of the study than would he.

A day or so later, Bill received a follow-up call from Meyers informing him that Meyers had spoken with NASA administrator Dan Goldin, during which he had repeated Bill's quote. Goldin's response was to the effect "That's exactly what we need. I want that guy – get 'em!" Again, Bill had been "had." Like it or not, he specifically had been pegged to critique the JSC findings. This was a job Bill did not want to do – it is not his style to be out there "front and center." Still, what choice did he have? Goldin was an appointee of two US presidents (and was later reappointed by a third). Bill acquiesced.

Bill then received a pre-publication copy of the soon-to-appear "Mars fossils paper" and realized that the JSC team had changed their story – now reporting exceedingly tiny bacterium-shaped fossil-like objects (Figure 8.8) – and also knew that the telling data they reported were scanty, so scanty in fact that had the paper dealt with ancient fossils on Earth, he was reasonably certain that it would not have been publishable.

Bill was later told that their submission to *Science* had received six reviews, evidently twice the number then normally requested – of which three were favorable and three were not – and that the journal's entire editorial board, some 12 members in all, had met to resolve this conundrum. And a conundrum it apparently was, because *Science*'s major international competitor, *Nature*, had reputedly agreed to publish the paper if *Science* did not.

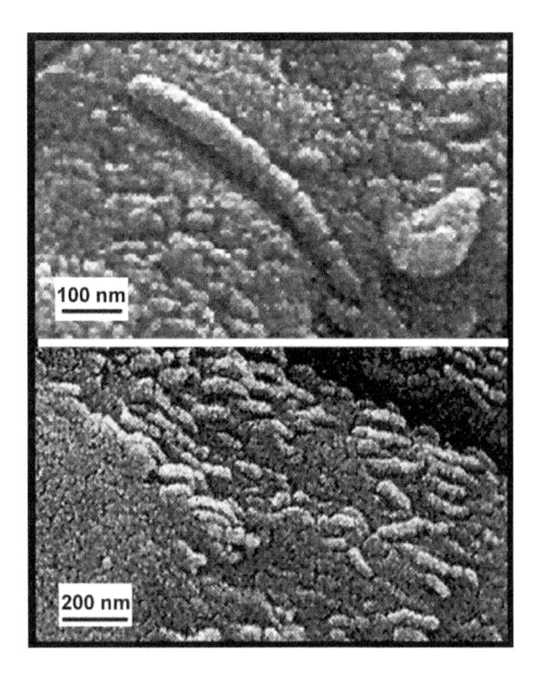

FIGURE 8.8 "Possible fossils" in ALH84001 unveiled at the "August 7th Event."

To an outsider, this bit of "blackmail" may be a bit difficult to fathom. Nevertheless, the story rings true, well illustrating the long-term battle for international supremacy between these two leading "hot topic" scientific journals, each committed to their own perceived needs (notoriety and revenue). Presumably to counter this threat, *Science* did in fact publish the paper, but to protect its reputation the journal editor wrote a skillful disclaimer at the preface of that issue indicating that, although this claim of past Mars life may or may not hold up, it was of such potential importance that it needed to be known and evaluated by the scientific community.

On Monday of that week in August 1996, Bill began to prepare his comments when he was then confronted by another unwelcome problem. The press conference at NASA headquarters that had originally been planned for Thursday of that week, shortly before the publication of the paper in *Science*, had now been rescheduled for Wednesday, a day earlier. (This was notably odd, but it was evidently spurred by Dick Morris, President Clinton's political strategist, who is reported to have revealed the news to his lady-friend, Sherry Rowlands, whom he had housed near the White House at Washington's Jefferson Hotel, startling findings that Ms. Rowlands in turn is said to have passed along to either a US aerospace magazine or the London tabloids or perhaps both.)

In any case, Bill flew to Washington that Tuesday and arrived at NASA headquarters early Wednesday morning for a pre-conference rehearsal. There were seven or eight principals scheduled to present remarks: NASA's Administrator Goldin was to lead off, seconded by Deputy Administrator Wes Huntress, with their presentations to be followed by those of the contributors to the paper and Bill's evaluation to come last. When his turn came at this run-through, Bill presented his evaluation. Because 1996 was an "Olympic Year" – and because he wanted his remarks to be understandable to the public – he rated various aspects of the report on a scale of 1 to 10, like an Olympic event. He gave the probability of the meteorite actually coming from Mars a score of "9"; gave the inferred 3.6-billion-age of the rock an "8"; and assigned the putative fossils as being evidence of life on Mars a score of "2," noting that to establish the existence of past life on Mars would require additional data.

The NASA Public Affairs officer at the rehearsal winced at Bill's scores and skeptical appraisal and urged the reporting scientists to be decidedly more positive. To Bill, at least, her criticism seemed out of place, directed primarily at Dave McKay, a fine person and a thoughtful, reserved scientist who was simply trying not to overstate, not to misrepresent his case. But, clearly, the Public Affairs officer, neither knowing the data nor understanding the workings of science, was putting on the pressure to make NASA "look good."

Subsequent accounts have reported that during this rehearsal, Dave McKay recounted an anecdote about having left a photo of one of the supposed "Mars fossils" for his 13-year-old daughter, Jill, to find and then asked her "What does it look like?" Her response is said to have been "in a youthful, know-it-all tone: Bacteria." Bill was present at the run-through and does not recall Dave McKay having said anything of the sort. Interestingly, however, what Bill *does* recall is that Everett Gibson at some point told him that *he* had left a photo on the kitchen table at his home and that the following morning, his wife identified the imaged object to be a bacterial filament. When Bill asked, "How did she know?" Gibson replied to the effect that "She was then studying for a degree in nursing and was familiar with bacteria."

It is unclear as to whether either or neither of these accounts is accurate. But what is certain is that neither story was presented at the press conference, presumably because the media would have regarded such identifications as embarrassingly amateurish. In fact, as was confirmed when the *Science* paper appeared the following week, identification of the objects as "bacteria" was fanciful.

The inorganic vein-filling globular objects McKay and Gibson showed Bill at JSC when he had earlier visited were a few tens of microns across and, therefore, OK for possible "fossil protozoans" – though they had no protozoan-like biologic features that Bill could discern. But the iconic images of the bacterium-like objects made available to the media at the press conference (and repeatedly shown subsequently worldwide) had no scale-bars to show the size of the objects. How large, or small, or tiny, or minute were they? The public had no way to know. As it turned out, when the *Science* paper was published days later, the world could see them to be exceedingly minuscule, some 20 to 40 *nano*meters broad, much too tiny to be like any organism known on Earth, bacterial or otherwise. In fact, they are more nearly the size of ribosomes, the minute (20 nanometer-sized) protein manufacturing bodies inside cells, of which hundreds of thousands typically occur in each human cell and some 20,000–70,000 inside a single-celled bacterium. In other words, they are the size of the smallest intracellular bodies *inside* a bacterium, not the size of a bacterium itself.

Following the rehearsal, the group trundled upstairs to the auditorium, which was packed, cheek by jowl, with perhaps 500 reporters. The place was filled to overflowing. The speakers assumed their places on the stage – and then the lights flickered, dimmed, and went out. Because the reporters' plugged-in electronic gear had overloaded the system, NASA's auditorium was flooded in blackness. Administrator Goldin and Bill retired to the speaker-presentation "green room" adjacent to the stage. Bill had not previously met Dan Goldin. The two were there alone. The electrical problem lasted about 20 minutes and during that period Goldin told Bill that he had spent "three full hours" personally "debriefing" (his term) the JSC scientists before he had divulged their findings to President Clinton (a link in the chain of events that explains the JSC scientists-to-Goldin-to-Clinton-to-Dick Morris-to-Sherry Rowlands-to-the news leak that led to the rescheduling of the news conference).

In retrospect, it is evident that Dan Goldin was a properly demanding, hard-driving, first-class administrator of NASA's vast hierarchy, and that his deficiency in initially sizing up the JSC report stemmed from the fact that he was a CEO engineer (which, it seems exactly appropriate for NASA, is an outfit that builds exceptionally excellent novel instruments but has engineering rather than science as its primary goal). Thus, Goldin's problem was that he lacked a fundamental knowledge of Biology, Paleontology, and the history of life, the bases of the supposed life-on-Mars report, and for that reason had evidently asked the wrong questions of the JSC scientists. Because of this, he had potentially embarrassed his agency, his president, his country, and himself.

In any case, during this green room chat – when Bill listened far more than he talked – Goldin seemed to think that Bill was probably OK. And he seemed to like Bill even more when Bill outlined the problems of the "Mars fossils" report at a Caltech Memorial Service for Carl Sagan (Goldin then asking Bill after his presentation whether the JSC scientists knew about the problems Bill discussed. They did.) Goldin later invited Bill to visit him at a home near UCLA, where he was sequestered ostensibly on vacation (during a national political convention that Bill presumed Goldin preferred not to attend) and appeared to be "trying Bill out" as a potential NASA headquarters' appointee. (Not Bill's cup of tea.) Then, even later, at a UCLA symposium in front of an audience of 550, Goldin publicly praised Bill for "*having saved NASA's reputation*."

More recently, after Goldin departed government service but had volunteered to help NASA solve a then-vexing problem, he called Bill to ask his advice. Goldin set the ground-rules for the conversation: "I will call you at home, not the university" (the implication being that Bill's phone conversations at UCLA might be automatically recorded – which, one supposes, might be true). Goldin called. The two talked. Bill did his best to help. Toward the end of the conversation, Goldin opined (with what Bill took to be well-intentioned if over-the-top, hyperbolic exaggeration): "Bill, you are the only scientist that I trust."

Let us now return to the NASA press conference – which was televised live for one-and one-half hours over International CNN. The event was introduced by President Bill Clinton, speaking from the South Lawn of the White House, who stated that "Today, rock 84001, formed more than 4 billion years ago as a part of the original crust of Mars, speaks to us across all those billions of years and millions of miles. It speaks of the possibility of life. If this discovery is confirmed, it will surely be one of the most stunning insights into our universe that science has ever uncovered. Its implications are as far-reaching and awe-inspiring as can be imagined." The president's glowingly hopeful introduction was followed by remarks from NASA administrator Dan Goldin; associate administrator Wes Huntress (who identified Bill Schopf as the "designated skeptic"); the principal co-authors of the soon-to-appear *Science* paper (McKay, Gibson, Zare, and others; Figure 8.9); and finally Bill evaluating the work.

Despite Huntress' earlier assertion, rather than regarding himself to be a "designated skeptic," Bill viewed his responsibility to be that of presenting an unbiased appraisal of the science, and for this reason prefaced his remarks by noting that he preferred his comments to be regarded "as part of a discussion rather than a debate." (In fact, Bill's private hope was that the interpretation of the JSC group would prove correct – this would have been a momentous discovery and a huge spur to the search for life on other worlds – but this was not the place for him to express his personal preferences.) Bill said his piece, stressing that this was "a fine piece of work on a complicated interdisciplinary problem that is not easy science." Nevertheless, of the hundreds of reporters in the auditorium (Figure 8.10), Bill's impression was that virtually all "wanted to believe" – save one, from the *New York Times* who seemed to understand the holes in the JSC presentation. But science is not based on wishful thinking. Science must be founded on firm facts. "Possibly," "perhaps," "maybe," and "feel-good solutions" are insufficient. "Good enough" is not good enough. The JSC team did not have a smoking gun. They knew it. Bill Schopf knew it.

During a commercial break following his remarks, Bill walked over to the CNN video crew to observe how they did their job. The next part of their coverage was to feature immediate reactions of the viewing public, and he was mildly amused to view the response of a middle-aged lady, speaking with a southern drawl, who summarily chastised his remarks, stating that she "just couldn't understand why that

FIGURE 8.9 NASA's "Aug. 7th Event"; left to right: Wes Huntress, Dan Goldin, Dave McKay.

FIGURE 8.10 Media coverage at the "August 7th Event," NASA HQ, Washington, D.C.

other scientist [Bill Schopf!] is such a 'Doubting Thomas' when simple common sense tells us that life has now been found on Mars." Her remarks were not broadcast – and, as Bill thought at the time, were probably not much different from those his aunts in Iowa and New Mexico would have expressed.

A day or two later, after Bill had returned to UCLA, he received quite a number of emails from scientists across the globe. Most seemed to think that he had done OK; one email, from André Brack, an origin-of-life expert in Orléans, France, even complimented him on the tie he had worn! But another, from Belgian Nobel Laureate Christian DeDuve, criticized Bill for not being "more harsh" in his carefully worded skeptical appraisal. (Still, Bill thinks now, as he thought then, that the JSC team had done useful work on an important problem. So, perhaps, he was "softer" than he might have been.) In any case, it was abundantly clear that this press conference had attracted an enormous viewership, worldwide.

A little more than a year after this "August 7th 1996 event" at NASA headquarters, John Bradley and his co-workers (including Everett Gibson and Kathie Thomas-Keprta, co-authors of the McKay et al. *Science* announcement paper) published a paper in *Nature* showing that the putative Martian "possible microfossils" reported from ALH84001 are in fact wholly inorganic, composed of calcium carbonate and the silicate mineral pyroxene (December 1997 *Nature* 390: 454; Figure 8.11). Importantly, they employed the same new JSC instrument (an environmental scanning electron microscope [ESEM]) used in the original work by McKay and Gibson. Instruments of this type are very expensive but are exceedingly powerful, increasing the obtainable magnification of the images of objects studied by more than 10-fold from the previous maximum typically obtained by use of scanning electron microscopy (SEM), about 25,000 times their actual size, to 300,000 times (with the theoretical limit being some 1,000,000 times). The instrument was evidently used at JSC to detect minute cracks in the wings of the shuttles

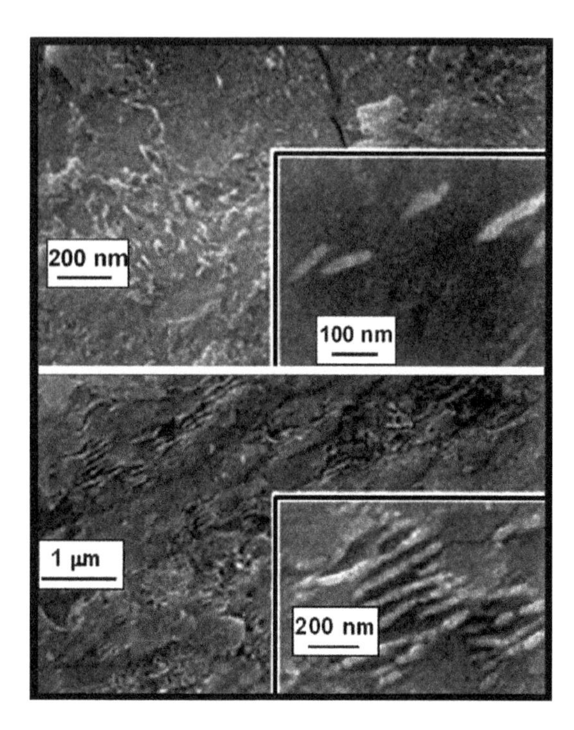

FIGURE 8.11 Carbonate (above) and pyroxene (below) pseudofossils in ALH84001 documented by Bradley et al. in 1997.

that transported astronauts and goods to the International Space Station, and as either McKay or Gibson later told Bill, they had been allocated a very few (only two or three) sessions from late at night into the wee hours of the morning to study the objects in ALH84001. But McKay and Gibson deserve credit. Their use of this new instrument to study objects in rocks was a "first," a pioneering effort to move science forward.

Pause for a moment and think about what a repeated 100–1,000-fold increase in obtainable magnification from visual to optical microscopy to SEM to ESEM actually means. Imagine that you are looking at the surface of a flat table. Increase your eyeball magnification by 100 times and you would see that the "flat" surface is covered by rises and pits – that's the highest magnification provided by optical microscopy. Then, up that magnification 25 times to 25,000 (the typical top magnification of SEMs, precursors of the ESEM) and you would see the rises and pits as hills and ravines. And, then, increase that magnification by more than another ten times, to 300,000, and the hills and ravines would now be mountains and deep valleys. Now, place yourself in the situation confronted by McKay and Gibson who evidently had no previous experience with the use of ESEM, an analytical technique new to the geosciences, and you can understand their plight (and the explanation for their misinterpretation of minute mineral lathes for orders-of-magnitude larger bacteria). They should not be faulted – they did the best they could. That's science.

Notably, however, the Bradley et al. 1997 report showed that the "possible Martian fossils" were solid, devoid of life-like cells, were common minerals (carbonate and pyroxene), and lacked organic carbonaceous components. Bill's skeptical evaluation of the far-too-tiny-for-life objects had been confirmed.

8.6 INTERLACING THE "UPS" AND "DOWNS"

Let us now fast-forward a decade to 2006 and tie together the previous two vignettes: Bill Schopf's 1993 report of discovery of especially ancient 3,465-million-year-old microbial fossils from Western Australia, and the 1996 "possible fossils from Mars" NASA headquarters event and the controversies they spawned. These two seemingly disparate stories were brought together in a 2006 book, *The Rock from Mars: A True Detective Story on Two Planets*, written by *Washington Post* journalist Kathy Sawyer, in which Bill is cast as the "fall-guy," an accomplished but difficult, less-than-admirable antagonist.

Initially, Ms. Sawyer's book presents Bill as a star: "noted paleobiologist ... heavyweight ... great scientist from California ... god of the Precambrian." But whether merited or not, by these seeming accolades, she also sets him up as a creditable adversary – the "devil's advocate ... fiery academic ... [an] intimidatingly forceful Goliath to McKay's David" – which makes for a better story, a more exciting read (though it is regrettable that because of Bill's supposed "unsparing rebuttal" of claims made at the NASA news conference, he is described as "McKay's nemesis" and is said to have become party to "bitter personal antagonism"). And it is unfortunately evident from author Sawyer's account that Bill's skeptical evaluation of the putative "Mars fossils" not only did not sit well with Dave McKay but also displeased his wife, who, after the Brasier-led criticism of Bill's work, is reputed to have begun signing her emails to friends, "With a faint glow of Schadenfreude."

It would not be appropriate to object to Ms. Sawyer's characterization of Bill Schopf – that's her business, not his – and it may well be that his style of presentation, evidently appreciated by the students in his classes, is in her view not suitable to a press conference setting. Nevertheless, it is true that Bill has never harbored even the slightest ill-will toward David McKay or Everett Gibson or anyone else whom he has evidently displeased. Moreover, there is absolutely nothing wrong with having any serious journalist such as Ms. Sawyer write about Bill and his work. After all, in his view, there are only four professions in our democratic society that he holds in greatest esteem – the judiciary, the professoriate, grade-school and high-school teachers, and journalists (the so-called "4th Estate") – chiefly because all of these can play a major role in helping the society to improve.

But what Bill does object to is that Ms. Sawyer, a no doubt highly regarded journalist, never took the time to interview him, to personally discuss with him the bases of his findings and assessments. That is wrong. Why? In Bill's estimation, the task of any serious journalist is to uncover the facts of a situation and the reasons underlying the principals' interpretations of those facts, considerations that he imagines too commonly hinge on personal self-serving predilections – "I'm right, I know my stuff, the other folks are wrong" – a deficiency that he chalks up to what he calls the "me-first gene." (To cut Ms. Sawyer some slack, it *is* true that she asked Bill to "fact-check" the lengthy excerpts of her book manuscript where he is mentioned, with which, because of her publication deadline and Bill's travel schedule, he had only three days to comply. He did so and to her credit, she then deleted several of her most grievous initial errors – even referring to Bill as "meticulous.") But it seems impossible for Ms. Sawyer or any other journalist to understand (or have even a reliable inkling) of the principals' motivations and in-depth points of view without personally interviewing all involved.

In Ms. Sawyer's acknowledgments in *The Rock from Mars,* she gives "Special Thanks" to 20 individuals, including Bill, "some" of whom she reports to have "participated in multiple lengthy interviews over a period of years." Her judicious use of the word *some* in preface to this phrase gets her off the hook. Her statement is not incorrect, for it is clear from her writing that she did interview many of the participants, though not Bill Schopf. So, she can be faulted for her journalism, just as can any academic who does not thoroughly ferret out the facts. Ms. Sawyer had the opportunity – she could have talked with Bill – but she did not. That is wrong!

To some, the acrimony recounted in Ms. Sawyer's book may seem a necessary essence of science. But many scientists would not agree. Indeed, and though some outsiders might argue that such unpleasantry is a necessary part of science, many of its practitioners think otherwise, it being far more preferable for scientists to sort out disagreements among themselves privately rather than burdening journalists or the public with their "unwashed linen" and thereby undermining respect for the basic goals and aims of the scientific enterprise. Moreover, though the penchant for highlighting such disagreements makes news and benefits the "bottom line" of the purveyors of the story, such battles of personality – a confusion of the messenger with the message – obscure the telling facts and are detrimental to science. Though they may be "entertaining," they do not help at all to properly address the crucial questions.

8.7 THE CURRENT STATUS OF THIS ANCIENT FOSSIL CONTROVERSY

Although Bill Schopf's initial 1993 report of the especially ancient (3,465-million-year-old) world's oldest diverse microbial fossil assemblage of the Apex chert garnered immediate widespread acceptance, during the first decade of this century, it spurred doubts, criticism, and alternative interpretations. Such questioning has continued to as recently as 2015 and 2016, when Professor Brasier's colleagues (with him being a posthumous co-author) yet again claimed the fossils to be mineralic artifacts, once again changing their tune – from "opaque minerals" to "abiotic graphite" to, in the latest instance, platy "phyllosilicate" and micaceous ("vermiculite-like") mineral flakes, the exteriors of which they imagined to have been coated by later-introduced crude oil-like fluids (an "add-on" necessitated by the need to rationalize the earlier documented carbonaceous composition of the fossils).

However, like the other claims that the Apex fossils are non-biological pseudofossils, this latest suggestion does not pass muster. The fossils are cylindrical, not platy, and rather than being solid and sheet-like, they exhibit three-dimensional cells enclosed on all sides by box-like organic walls, partitions that completely cut through the tube-like filaments to define the interiors of their regularly spaced cells (see Figure 9.10).

Moreover, analyses of originally described Apex specimens using the then-newly applied technique of confocal laser scanning microscopy (CLSM) show that, instead of producing fossil-like filaments, oil permeating into the Apex chert produces a three-dimensional "chicken-wire" fabric where the oil infills boundaries between the interlocking mosaic of tiny irregularly shaped quartz grains in which the fossils are permineralized (Figure 8.12). (Interestingly, this discovery was serendipitous, made during Bill's restudy of petrographic thin section–embedded specimens that he had examined years earlier using high magnification oil-immersion optical microscopy. The oil that he had then used permeated into the rock, infilling chert grain boundaries in the upper few microns of the section surface, and turned out to fluoresce brightly under CLSM laser light. He has since switched lab-wide to use of fluorescence-free microscopy immersion oil.)

And, finally, like the earlier proposed suggestions that the Apex fossils are composed of barium carbonate or abiotically produced "graphite," oil-coated mineralic filamentous fossil "look-alikes" are unknown in the geological record.

Thus, the Apex fossils are three-dimensionally composed of carbonaceous matter, not of organic-coated minerals, and not only are their cell walls composed of carbonaceous kerogen, but as Bill and his colleagues at the University of Wisconsin showed in 2018 – by analyzing fossil specimens in a slice of the very same chunk of rock in which the fossils were first found – the mix of the stable isotopes of carbon

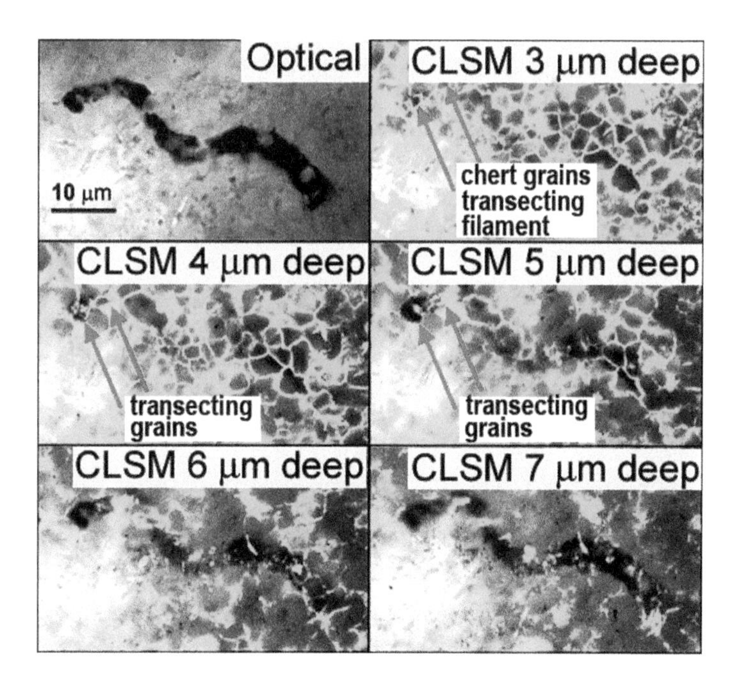

FIGURE 8.12 Apex filament showing "chicken-wire" fabric produced by rock-permeating oil.

(C-12 and C-13) in their carbonaceous matter also varies systematically from taxon to taxon. Of the five different Apex species analyzed (in 11 individual specimens), two have $^{12}C/^{13}C$ ratios typical of non-oxygen-producing primitive photosynthetic bacteria; one, a composition characteristic of early-evolved methane-producing Archaeal microbes; and two others that have values distinctive of methane-consuming microorganisms.

The recently proposed scenario of mineral-coating by permeating oil-like fluids is untenable. Such in-seeping oil could not and cannot "pick and choose" among co-existing microscopic objects in a rock on which to paint such differing and physiology-defining isotopic signatures while miraculously not coating adjacent objects with the same distinctive ooze. This must be especially true for fossils like those of the Apex chert, where the differing signatures are exhibited by specimens close-packed, only a few microns distant from one another, within sub-millimeter-sized granules. And even were such a scenario physically possible – which it is not – it strains credulity to imagine that the three isotopically differing seeping oils required preferentially implanted their physiologically definitive isotopic signatures on the surfaces of multiple specimens of five different tiny fossil species identified solely on the bases of their cell size and overall morphology, not their isotopic compositions.

Thus, not only are the Apex fossils assuredly *bona fide* but the firm facts of this 2018 study now also reveal how they lived. Nevertheless, it remains true that these especially ancient minute cellularly preserved carbonaceous microbes have now for more than a decade withstood considerable doubt and skepticism – as it seems likely they will for all of time. How could this be? How could this find weather such a storm? The answer is fairly simple. Answers to scientific questions must be based on facts, not personal predilections, the task of science being to describe the real world as thoroughly and honestly as it can and let the chips fall where they may. For the Apex fossils, the following are a few of the most telling facts:

1. The fossils are numerous and diverse (175 originally described specimens assigned to 11 taxonomic categories, now upped to 193 specimens) – the point being that microorganisms live in multicomponent communities, and that if one member of a community can be preserved, others should be also.
2. The 3,465-million-year-old Apex microbial assemblage does not stand alone. Hundreds of other Precambrian units contain similarly preserved microbial fossils, including nearly a dozen occurrences of microfossils and stromatolites reported from geological units of nearly the same age and appreciably more in other rocks that are only a few hundred million years younger.
3. The Apex fossils are demonstrably composed of coal-like carbonaceous organic matter – shown most convincingly by Raman spectroscopy – as are the literally millions of such fossils now known throughout the Precambrian geological record.
4. Detailed in-depth analyses of the chemistry of the Apex organic matter show it to be unquestionably of biological origin.
5. Like their modern microbial counterparts, the Apex fossils are sinuous garden hose–like tubes composed of disc-shaped, quadrate, or elongate

cells – some 1,900 cells being measured initially, a sum now upped with additional specimens to well over 2,000 – shown by optical microscopy, three-dimensional Raman imagery, and confocal laser scanning microscopy (analytical techniques that are outlined in the following chapter).

6. And most recently, in January 2018, Bill's colleagues and he used secondary ion mass spectrometry (SIMS, yet another technique outlined in the next chapter) to document the physiology of Apex microbes, the way they lived. To assure themselves that their results would hold up, they analyzed 11 different specimens of five Apex species preserved in the same rock specimen from which the fossils were first described in 1993 (an exceptionally arduous, time-consuming task).

   These SIMS results are particularly telling, showing that this exceptionally ancient microbial community was notably diverse including primitive photosynthetic members of the Bacterial Domain of the Tree of Life, early-evolved methane-producing members of the Archaeal Domain – the first Archaea known from the geologically ancient Archean Eon (or from rocks of any other age) – as well as methane-using microbes.

   The Apex methane producers ("methanogens") are assuredly Archaea, this being the only biologic group known to generate methane gas. And the Apex methane consumers ("methanotrophs") are either similarly early-evolving methane-using Archaea or are primitive members of the Bacteria Domain (technically, γ-Proteobacteria), these being the two biological groups capable of using methane gas as the source of the carbon needed to build their cells. Moreover, the carbon isotope signatures (the ratio of C-12 to C-13 expressed on the standardized scale used in such studies as $\delta^{13}C$ in parts per thousand, ‰) of the various groups are unlike oxygen-producing cyanobacteria (which have a $\delta^{13}C$ value of about –25‰) and distinctly differ one from another – photosynthetic bacteria having a $\delta^{13}C$ value of about –30‰, methanogens of about –35‰, and methanotrophs of about –40‰. These differences are sufficiently great that the life style–identifying carbon isotope signatures of the groups cannot be readily confused or misinterpreted – and all differ distinctly from the –27‰ value of non-fossil associated particulate kerogen dispersed in the Apex chert, measured both by SIMS in this study on individual kerogen flakes and previously by the PPRG on kerogen isolated from large bulk samples. Importantly, the physiologically distinctive fossil-determined values also establish that all three of these evolutionary lineages, know from molecular biologic studies of living microbes to fit near the very base of the Tree of Life, were present early in Earth history. Indeed, this is exactly the composition of the Apex biota that knowledgeable microbiologists had hoped for and expected.

   In sum, the facts quite clearly support the major conclusion of the 2018 study, namely "that methane cycling methanogen-methanotroph communities were a significant component of the Paleoarchean biosphere, [a finding that given] the oxygen-deficient setting inhabited by extant methanogen-methanotroph communities is consonant also with an anoxic early environment."

The Archaeal methane producers and methane users of the 3,465-million-year-old Apex assemblage occur together in what is known as a "microbial consortium," a type of naturally occurring association of two differing types of co-existing interactive microbes. In this instance, one of the members of the consortium produced methane gas (CH_4) that was then used as a nutrient by another. Such mutual interactions are exceedingly common in modern microbial communities, with the Apex consortium being broadly similar to another fossilized consortium that Bill and his colleagues reported in 2017 from rocks 3,400 million years old, in which sulfate-reducing bacteria produced hydrogen sulfide gas (H_2S) that was used as fuel by co-existing photosynthetic bacteria.

7. Thus, if the 3,465-million-year-old Apex fossils and this only slightly younger 3,400-million-year-old sulfur-cycling community are considered together, it is evident that Earth's earliest known biota included at least four distinct primitive microbial lineages – sulfate reducers, methane producers, methane consumers, and anoxygenic photosynthesizers – all inhabiting an oxygen-deficient primordial world and all representing lineages that fit near the very base of the rRNA Tree of Life (Figure 8.13).

The evidence is sound. Not only is the nearly 3,500-million-year-old Apex assemblage composed of authentic microscopic fossils, but by this time the two most

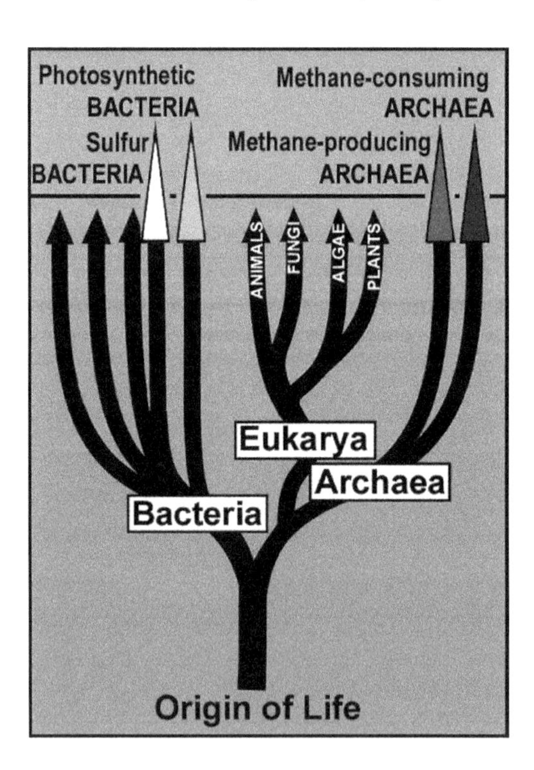

FIGURE 8.13 Tree of Life showing four lineages older than 3,400 million years.

ancient Domains of Life, the Archaea and the Bacteria, had already split apart, indicating that the origin of living systems must have occurred substantially earlier. How much earlier is yet unknown, although there are hints of life – a few carbon isotope measurements of graphite, one report of a putative stromatolite, and a finding of fine sinuous tubes in volcanic rocks that might have been formed by microorganisms – dating from 3,800 to as early as 4,200 million years ago. *Clearly, primordial life evolved earlier, farther, and faster than had ever been imagined!*

8.8 DISPARATE PHILOSOPHIES OF SCIENCE

For those interested in how this paradigm-changing science has progressed – and more generally, how science is done – it is useful to compare the differing styles of science that are illustrated by the Apex fossil controversy.

On the one hand, there is the *fact-based cascade of evidence approach*, the one Bill Schopf favors, the notion being that if a particular interpretation is correct, then additional supporting evidence must also be present, then more and then still more, a cascade of mutually reinforcing clear-cut facts each testing the plausibility of the preceding lines of evidence and building confidence in the conclusion drawn – in this case, that the Apex assemblage is in actuality composed of *bona fide* fossils.

The other approach, spurred by the writings of the philosopher Karl R. Popper (1902–1994), stems from his "Principle of Falsification," in essence the idea that a scientific interpretation should not be accepted unless all plausible alternative explanations can be rejected.

Popper's *plausible alternative hypothesis approach* is excellent, absolutely meritorious – consideration of alternative null hypotheses is a hallmark of sound science. Rather, as the Apex fossil controversy well illustrates, the crux of the problem lies in what one assumes to constitute a "plausible" alternative explanation – simply because what seems obvious to one worker may not seem so to another. For example, one worker might propose what in that scientist's opinion is a "plausible" alternative hypothesis to a previous interpretation, a notion that the initial proposer of the concept regards as unsubstantiated speculation.

Such differences of opinion commonly reflect differing levels of knowledge and experience, and can ultimately be resolved only on the basis of established facts. Thus, for example, to a knowledgeable worker, the dark-brown color of the Apex fossils suggests that they are composed of geochemically mature carbonaceous matter like that known from similarly ancient partially "pressure-cooked" fossiliferous deposits, whereas to a less experienced worker, their dark color might plausibly suggest that they are aggregates of opaque minerals. Similarly, an experienced worker would be well aware that the occurrence of *non*-biologically produced coaly kerogen or redefined "graphite" pseudofossils, barium carbonate fossil "look-alikes," or fossil-mimicking chain-like aggregates of crude oil–coated mineral grains are not only implausible but also wholly unknown in the geological record – whereas to those lacking the requisite background, such interpretations of the Apex fossils might seem perfectly reasonable, in Popperian terms, "plausible." And the error of misinterpreting secondarily emplaced mineralic veins as authentic microscopic fossils, long known and well documented in the paleobiological literature, is yet another mistake that experienced workers would studiously avoid.

In short, it is not the Popperian alternative hypothesis approach *per se* that has led some to erroneously question the biological origin of the Apex fossils; it is the misapplication of Karl Popper's central thesis and the definition of what constitutes a "plausible" null hypothesis. Indeed, properly applied, the Popperian *alternative hypothesis* strategy and the *fact-based cascade of evidence* approach fit together hand-in-glove. But their normally expected consonance failed (in retrospect, quite notably) in the onslaught of criticism of the Apex fossils.

To highlight the differences between the two approaches, Table 8.1 presents a side-by-side synopsis outlining how each unfolded from 1993 to the present.

As is shown in Table 8.1, the facts indicate that the Apex fossils are not composed of opaque minerals, are not composed of non-biological "graphite," are not crude oil–coated mineral grains, and are not mineral crystallites, needles, or vein-fillings. In Popperian terms, each of the "alternative hypotheses" has been rejected.

In retrospect, however, it seems striking that as each such alternative explanation has been shown to be mistaken, the justification for the *non*-biogenic interpretation has been repeatedly revamped to form a new pathway to the same conclusion. These oft-repeated changes give the impression that rather than being a dispassionately constructed Popperian *alternative hypothesis*, the conclusion presented by those advocating a non-biological origin of Apex fossils has been a preconceived notion,

TABLE 8.1
Comparison of Differing Interpretations from 1993 to the Present of the 3,465-Million-Year-Old Apex Fossils

Fact-Based Cascade of Evidence	Alternative Hypotheses
1993: Numerous fossils are indigenous and syngenetic to the Apex chert, interpreted by optical microscopy to be composed of organic matter.	2001: Yes, the described Apex fossils are numerous, indigenous, and syngenetic; but no, so-called fossils are composed of opaque minerals.
2002: Raman shows the fossils are composed of biogenic carbonaceous kerogen as are all other permineralized Precambrian and younger organic-walled fossils.	2002: Yes, the Apex putative fossils are carbonaceous. But no, they are composed of a *non*-biologically formed graphite-like mineral, not biogenic kerogen.
2005–2015: CLSM studies confirm that the Apex fossils are composed of kerogen. Detailed chemical analyses show that the kerogen is biogenic. Optical microscopy, CLSM, and 3-D Raman show that the fossils are cylindrical and composed of cells. CLSM shows that permeating oil produces a 3-D chicken-wire mosaic, not carbonaceous filaments.	2005–2016: Yes, the carbonaceous matter is biogenic. But no, the fossil-like objects may be barium carbonate crystallites, needles of silica, planar quartz-hematite vein-fillings, or platy minerals coated by seeping oil.
2018: Isotope measurements of individual fossils show that the five Apex taxa analyzed fit near the base of the Tree of Life.	

its underpinnings repeatedly altered to fit the new-found facts. If this is true it is perhaps fair to wonder whether such an approach constitutes sound science (and, indeed, to wonder whether Karl Popper would have approved).

At the same time it is instructive to learn from this saga how the two seemingly disparate approaches – the "fact-based cascade of evidence approach" and the Popperian-based approach of "plausible alternative interpretations" – more typically fit together seamlessly, "hand in glove."

The initial in-depth report of the Apex assemblage showed the morphologically identified fossils to be numerous, indigenous, and syngenetic to the Apex chert, but because of a lack of appropriate analytical techniques, could only infer their carbonaceous composition (on the bases of their optically discernable color, texture, and similarity to previously known Precambrian and modern microorganisms). Such uncertainty about their composition, however, was soon laid to rest by Raman analyses of the originally described specimens, which established them to be composed of coal-like carbonaceous kerogen, a finding confirmed by their CLSM-determined fluorescence.

But was this kerogen demonstrably biogenic? Yes, as was then shown by detailed chemical analyses. Yet to be considered *bona fide* filamentous fossils, the structures should be cylindrical and exhibit discernable cells and cell luminae – all of which were then soon documented by 3-D Raman and CLSM. And finally, what types of microorganisms were these, how did they live? This final piece of the puzzle was put in place by SIMS analyses of fossils permineralized in the same chunk of rock from which the fossils were first described, showing that the Apex assemblage includes photosynthetic bacteria and methane-producing and -consuming microbes, ancient tiny forms of life that fit near the base of the Tree of Life.

Each of the tests in this cascade of evidence could have turned out differently. That is, in Popperian terms, results of each of the tests could have shown the interpretation to be in error, to be "rejected." Optical microscopy might have shown the fossils to be few, not numerous; or to be modern contaminants not indigenous to the chert; or to be secondarily emplaced, not syngenetic with the fossil-embedding rock. Raman might have shown them to be mineralic, not carbonaceous. Chemical analyses might have shown their carbonaceous matter to differ from preserved biogenic organics. Three-dimensional Raman and CLSM might have shown the fossils to be planar, not cylindrical; or solid and not composed of cells. And SIMS might have shown their isotopic composition not to correlate with their size- and shape-defined defined species or to be unlike the physiologies of modern microorganisms.

The findings of each such Popperian "alternative" null hypothesis could have cast doubt on the biological origin of the Apex fossils. But the fossils passed each and every test. To most scientists, this test-and-test-and-test-again approach constitutes sound science, slowly but surely building confidence in the resulting conclusion.

As the foregoing discussion clearly illustrates, doing science that "matters," asking new questions, and making new discoveries that move science forward is a difficult, commonly complicated, prolonged task. And, if you stop to think about it, the very fact that there has been such a long-term dispute about the Apex fossils shows that they actually *do* "matter" (after all, if they didn't, no one would bother to worry about them!)

Moreover – and importantly to those not fully acquainted with the workings of science – the discussion also shows that those who play this game are just like everyone else. All involved have preconceived notions – like every single one of us – derived from our parents, or our religion, or our culture, or some other source. And we each are imbued with the "me-first gene" necessary to maintain our own self-esteem, a feeling of self-worth (and its pre-programmed side-kick, hubris). But as the history of the Apex fossil controversy shows, when all is said and done – or, if you prefer, "at the end of the day" – feel-good solutions and preferred conclusions do not matter. *Firm facts should and will always win.* That is the way we humans succeed in understanding the natural world – and that is the way science will continue to progress for all of time!

Data now available about the Apex fossils – results of the most rigorous, in-depth, detailed studies of any fossilized microbial assemblage in the history of science – establish the current "gold standard" for Precambrian Paleobiology. Because of the time-after-time-after-time strategy of seeking answers to a cascade of questions, this long-term exercise exemplifies an effective way for science and knowledge to progress. Moreover, the result of these studies has now provided a linchpin to the solution to Darwin's quandary about life's earliest history, showing it to date from the earliest epochs of Earth's existence – a result that will please virtually all who have addressed the problem. And though Darwin could not possibly have imagined that his great dilemma would finally be put to rest 150 years after his *Origin* appeared, one would like to think that he, too, would be pleased.

9 Today's Status of This Science

9.1 MAJOR PLAYERS IN THE SCIENCE

In the years following the fundamental breakthrough reports of the 1960s by Timofeev, Tyler, Barghoorn, and Glaessner – the founding great leaders in the search for evidence of Precambrian life – many workers in many countries have come to the fore. Prominent among these have been Glaessner's star student, Precambrian paleobiologist and stromatolite expert Malcolm Walter, late Precambrian metazoan authority Jim Gehling, acritarch specialist Kath Grey and, in recent years, Archean fossil researcher David Wacey, Australia's leading experts in these aspects of the field; Precambrian paleobiologists and geologists Hans Hofmann and Guy Narbonne, Canada's foremost contributors; paleontologist Stefan Bengtson and acritarch specialists Gonzalo Vidal and Malgosia Moczydłoska in Sweden; Precambrian paleobiologist Nick Butterfield and Archean life researcher Martin Brasier in England; paleobiologist-astrobiologist Frances Westall in France; paleobiological electron microscopy expert Emmanuelle Javaux in Belgium; Precambrian fossil expert Thomas Fairchild and his students, Luana Morais and Cléber Atabike, in Brazil; geologist-paleobiologist Wlady Altermann in South Africa; and Archean fossil experts Ken Sugitani and Yuichiro Uena in Japan.

Russian, Indian and Chinese scientists have also been major contributors, including such notables as late Precambrian (Vendian) metazoan experts Boris Sokolov and Mikhail Fedonkin, and Precambrian stromatolite-microfossil authorities Igor Krylov, Volodya Sergeev, Victor Podkovyrov and Konstantin Nagovitsin in Russia; prescient paleobotanists-paleobiologists M.C. Mishra, B.S. Venkatachala, and Mukund Sharma in India; and, in China, such leaders as Yin Leiming, Hsu Jen, Li Xing-Xue, Cao Ruiji, and Ouyang Shu.

A great many scientists in the United States have also made splendid contributions, most prominently Barghoorn's star student and polymath Andrew Knoll, late Precambrian metazoan expert Dave Bottjer, Precambrian paleobiologists Shuhai Xiao, Andy Czaja, Dorothy Oehler, Chris House, Maud Walsh, Susannah Porter, Phoebe Cohen, Roger Buick, and Jeff Osterhout; pioneering organic geochemist John Hayes; molecular biomarker trailblazer Roger Summons; Raman specialists Anatoliy Kudryavtsev, Jill Pasteris, Craig Marshall, and Andrew Steele; isotope geochemists John Valley, Kevin McKeegan, Ken Williford, Mark Harrison, and Elizabeth Bell; and astrobiologists-geologists-stromatolite specialists John Grotzinger, Abigail Allwood, and David Flannery.

Many others also have played similarly crucial roles, primarily with regard to the interrelated co-evolution of environment and life in the ancient Precambrian world, including such notable contributors as isotope geochemists Jan Veizer in

Canada, Yanan Shen in China, and Françoise Robert in France; carbon isotope geochemist Manfred Schidlowski and geochemist and sulfur isotope specialist Harald Strauss in Germany; and field geologists Nic Beukes in South Africa and Arthur Hickman and Martin Van Kranendonk in Australia. In the United States, such "movers and shakers" include geologist-geochemist-environmental-evolution savant H.D. "Dick" Holland; Precambrian geologist-sedimentologist Don Lowe; Precambrian paleosol specialist Greg Retallack; noted mineralogists and banded iron formation experts Kase Klein and Clark Johnson; organic geochemists Ian Kaplan and George Claypool; astrobiologist-microbiologist-organic geochemist David Des Marais; atmospheric evolution gurus James C.G. Walker and Jim Kasting; and paleoenvironmental isotope geochemists Mark Thiemens, James Farquhar, and Paul Knauth. And the list of major players could easily go on and on: microbiologists whose studies of the modern world have contributed markedly to understanding the ecology and biological affinities of ancient Precambrian microbial communities; molecular phylogenists who routinely use the fossil record to calibrate their evolutionary trees and whose insights have time and again proven instructive to the field; and the growing number of molecular biologists using the technique of ancestral sequence reconstruction (ASR) to unravel the abbreviated history of life encoded in the genomes of living organisms – perhaps foremost among them Akihiko Yamagishi in Japan and Eric Gaucher in the United States – a increasingly productive "growth industry" that can be predicted to pay great dividends in the coming decades.

9.2 CONTRIBUTIONS OF THE SUBDISCIPLINES

These and many others as well, some the students either of the original field-founders or of the more recent major contributors, are advancing this science at breakneck speed. All of these workers deserve their well-earned credit. Indeed, every one of the diverse areas of expertise represented by these many workers provides a piece of the puzzle needed to unravel the mystery of life's early history. Thus, for example, the discovery of fossils of sub-Cambrian metazoans (multicellular animals) extends the known record of animal life into the uppermost reaches of Precambrian time. The stromatolites of the Precambrian show that light-using photosynthesizing microorganisms were widespread throughout photic-zone environments in which the numerous stromatolitic forms reveal the nature of the local physical environment (relative depth, wave action, and so forth). Studies of acritarchs (floating microscopic algae) document the planktonic floras of Precambrian seas; analyses of the chemical structure of molecular biomarkers (preserved organic molecules) can reveal their biological source; and measurements of the stable isotopes of sulfur in biogenic materials can indicate the presence of sulfur-metabolizing microbes, whereas those of carbon can establish the presence of light using photosynthesizers and methane-producing and methane-consuming microorganisms.

Nevertheless, applied individually to the cardinal question posed by Darwin, the "inexplicable … absence of vast piles of strata rich in fossils beneath the Cambrian system" – the paradigm shift from an unknown to a known record of early life set in place in 1965 – each of these approaches, in and of itself, falls short.

The early 1960s announcement of the findings of latest Precambrian metazoans extended but did not fundamentally alter the history of life as it had previously been known. Similarly, the occurrence of stromatolites, by themselves, could not answer Darwin's question. Indeed, the Cambrian-age specimens at Saratoga Springs were first noted by John Steele in 1825, and comparable examples were reported by Charles Walcott from Precambrian rocks of the Grand Canyon in 1883 and of the Lewis Range of Montana in 1902 to 1907, but their biological origin was not was not widely accepted until modern analogues were discovered by Australian geologist Philip Playford at Hamelin Pool on the western coast of Australia in the early 1960s. And here it was not the occurrence of stromatolites *per se* that established their biogenicity but, rather, the identification of the mat-building microorganisms (cyanobacteria) that promoted their formation – a finding soon extended into the Precambrian by the 1965 reports of the stromatolitic microbial communities of the Gunflint and Bitter Springs cherts. Thus, it was the discovery of the cellular stromatolite-forming microbes that proved the case, not simply the existence of stromatolites.

Planktonic Precambrian acritarchs by themselves might have solved the problem, but their misidentification by Walcott in 1899 as compressed remnants of shelled animals, their typically flattened distorted preservation, and the dearth of follow-up studies until the mid-1950s, when they were again misinterpreted – this time by Timofeev as the spores of geologically far younger land plants – raised questions as to their relevance and authenticity (with the distrust of Timofeev's reports being magnified by the Cold War politics of the day).

And considered alone, the other lines of evidence are also insufficient. Analyses of molecular biomarkers by themselves can provide only an ambiguous solution to Darwin's central quandary, primarily because of the notorious difficulty of establishing that the detected molecules date from the time of deposition of their host rock rather than being contaminants emplaced later either by seeping groundwater or during sampling or processing. And biogenic interpretations of the stable isotopes in Precambrian sediments, dating from the 1950s post-World War II period and the increased availability of commercially produced mass spectrometers, were initially understandably greeted with widespread skepticism, chiefly because they lacked direct confirming fossil evidence of the organisms that had generated the so-called "biogenic" isotopic signals.

In fact, of all of the diverse lines of evidence of Precambrian life, the only one capable of being sufficient in and of itself to establish beyond doubt the so-called "missing" pre-trilobite, Precambrian history of life is that identified by Darwin himself: "strata rich in fossils." Thus, as Darwin had presciently implied, the telling data were ultimately provided by the discovery of fossils – not those of more readily detectable large early-evolved animals, as he and the global scientific community had presumed for a full century following publication of the *Origin*, but of three-dimensional, cellularly preserved microscopic organisms. It is for this reason that the 1965 seminal studies of the diverse microbial communities of the Precambrian Gunflint and Bitter Springs deposits stand out as providing the long-sought "smoking gun" needed to solve Darwin's dilemma.

Moreover, it was important that at the beginnings of the modern science, there were two pivotal back-to-back reports of such ancient cellular microbes. The first, that of the Gunflint fossils, was the seminal "big find," the breakthrough discovery – but the fossils it recorded were marooned in time, organisms claimed to be an unbelievable billion-and-a-half years older than accepted dogma permitted. In fact, the "known" absence of a Precambrian was so deeply entrenched, handed down time and time again from teacher to student and recounted in textbooks and museums worldwide for a full century – and because the Gunflint fossils so obviously flew in the face of that universally accepted "fact" – that despite Cloud's independent validation, they were widely regarded as suspect, some sort of inexplicable major mistake (a fear that Barghoorn had long harbored and, as he told Bill, one of the reasons he had delayed so long in formally describing the Gunflint fossils, a process that he knew would enshrine them in the literature for all of time).

Thus, Barghoorn was personally much relieved when Bill had earlier shown him the results of his first studies of the Precambrian fossils of the central Australian Bitter Springs chert (rock samples that Barghoorn had collected a year earlier, alerted to this locality by field geologist Helmut Wopfner, whom he had met by chance in a saloon in Adelaide, and upon his return had turned over to Bill to examine). In short, when the unwelcome surprise of the Cloud manuscript describing his own studies of the Schreiber Beach Gunflint fossils arrived for Barghoorn's evaluation, he already knew that he was "in the clear" – that a second follow-up report would soon prove the case.

The second seminal report, published a few months later and recording the occurrence of the geologically younger Bitter Springs cellular fossils – better preserved than their Gunflint counterparts and preserved in different rocks of a different age from a different continent – was the "deal sealer," showing that a record of early life not only exists but is also, in fact, both widespread and decipherable. Taken together, it is these two discoveries that moved the science ahead and set it on its current course.

Such a "first find" and second "confirming discovery" are not atypical of paradigm shifts in science. The first find sticks out like a sore thumb: "It's gotta be wrong. I've never heard that before." The second find proves the case, showing that the first was not some one-off, misinterpreted fluke, as the scientific community turns from skepticism, to grudging acquiescence, to accepted fact. Here, the paradigm changed within a scant six months, almost overnight. But without a second confirming find, the first would have been ensnared in doubt and uncertainty. This find and find again is the way of science, the way confidence is built in its quest to understand and describe reality.

9.3 HOW FOSSILS ARE PRESERVED

As in many fields of science, the marked recent progress in studies of three-dimensionally cellularly preserved Precambrian microscopic organisms has been spurred by the development of new analytical techniques, the four currently most widely adopted such new techniques – confocal laser spectroscopy microscopy (CLSM), Raman spectroscopy, fluorescence spectroscopy, and secondary ion mass spectrometry (SIMS) – having been introduced to Precambrian Paleobiology by the work of Bill Schopf and his students. Interestingly, the need for several of these advances was apparent to Bill even in his mid-1960 graduate-school days, after which – at many times and in many ways – he set out to address them. He failed repeatedly until the last decade or so, when these efforts were finally rewarded.

The very best, most "life-like" three-dimensional cell-containing fossils known to science are those preserved by permineralization (known also as *petrifaction*, as in the common term "petrified logs"), a process in which the cell walls of a preserved organism become infused before they decay by permeating mineral-charged ground waters that fill in the microscopic watery spaces in their robust organic walls. This mélange then solidifies into a firm mix of organic wall and mineral, most commonly the mineral quartz that makes up the rock type known as chert, but also the minerals calcite, apatite, and gypsum.

To understand the process, do the following "thought experiment": Take a rose, place it in an ice-cube tray, and freeze it in a block of ice (Figure 9.1). What do you have? Not just a "really cold rose" but a frozen, ice-embedded rose, permineralized in the mineral ice. And, surprising as it may be, ice meets the definition of a mineral – it is naturally occurring and has both a definitive chemical formula (H_2O) and a characteristic (hexagonal) crystal form. (All this, one can guess, makes snow a sediment and compacted frozen snow a rock!)

Even without actually doing this experiment, you know that the veins and cells of the ice-permineralized leaves of the rose would still be intact, easy to see (Figure 9.2).

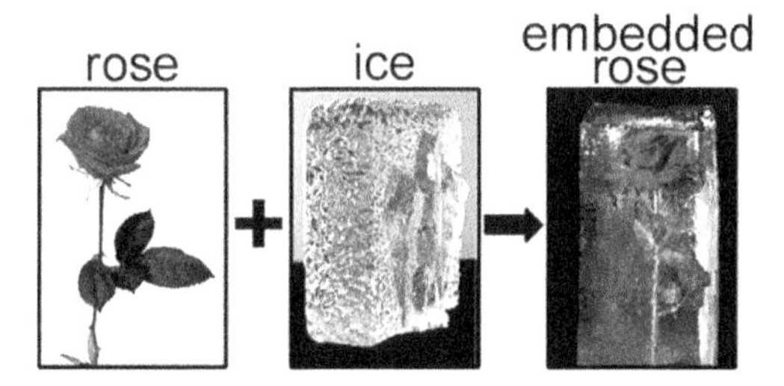

FIGURE 9.1 Artificial "ice-permineralization" of a rose.

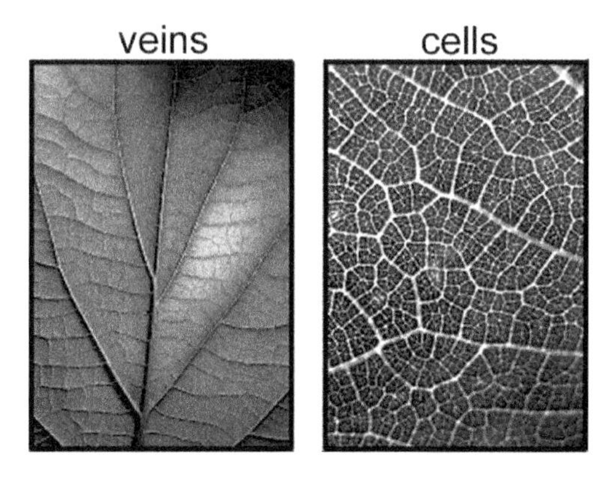

FIGURE 9.2 Veins and cells after artificial "ice-permineralization."

The permineralizing fluid (water) would have flushed away the cellular juices (the cytoplasm) of the interiors of the cells, infilling not only the cell cavities (the cell lumina) but also the tiny watery spaces in their surrounding cellulosic walls, where it would have then solidified into a firm mix of cellulose and ice without disrupting the three-dimensional size and shape of the cells.

Because they have robust shape-defining cell walls, such mineral permineralization occurs in plants, algae, fungi, bacteria, and Archaea – but never in the tissues of animals, in which the cells are enclosed in a flexible, albeit strong, membranous layer. Think about it. Chew a piece of celery. You hear it crunch. Your molars are smashing the robust organic cell walls. Now do the same with a bite of steak, or of fish, or better still an oyster (the meat, not the shell!) You don't hear any crunching – there are no robust cell walls in animals to smash. (As Bill jokingly tells the students, "*Animal cells have a low Index of Crunchablity.*")

Virtually all Precambrian cellular fossils are microscopic and preserved by one or the other of only two ways: compression or permineralization. The compression fossils typically occur in fine-grained shales, lithified beds of silt where their remains have been preserved between layers of accumulating sediment. As a result, as the overlying sediment builds up, they become increasingly flattened, two-dimensional, and their original shape and form becomes increasingly squashed and distorted. Permineralized fossils are different. Though less common than their compression-preserved counterparts, their remains are preserved in their original life-like three-dimensional form, as are all other microbes in their vicinity, a type of preservation that permits understanding of the complete ecosystem in their local environs. The paradigm-changing publications of 1965 on the Precambrian Gunflint and Bitter Springs fossils were based wholly on studies of life-like permineralized, rather than compression-preserved, fossils, a telling factor in their ultimate universal acceptance.

9.4 NEW ANALYTICAL TECHNIQUES: CONFOCAL LASER SCANNING MICROSCOPY

At the beginning of Bill Schopf's graduate-school days, publishable optical photomicrographs of the exceedingly minute rock-embedded cylindrical filamentous microbes that dominate the Precambrian biota were far from what was desirable. Because they required use of the highest magnification optical systems available (viz., 100x oil-immersion objectives), they could provide a useful image only of a thin in-focus plane and were restricted to specimens oriented parallel to the surface of a glass-mounted thin rock slice (a "petrographic thin section"), in which an entire specimen could be shown. As Bill had learned from his undergraduate Honors project, however, virtually all such fossils are sinuously bent or curved, wending into and out of any single focal plane like tiny twisted garden hoses (Figure 9.3). To defeat this problem, he devised the then-new technique – immediately adopted by workers worldwide and now in use for 50 years – of pasting together a front-to-back, one-after-the-other series of in-focus images of parts of such sinuous minuscule microbes acquired at sequentially varying focal depths. To these he later added interpretive drawings to better show their true three-dimensional overall form (Figure 9.4).

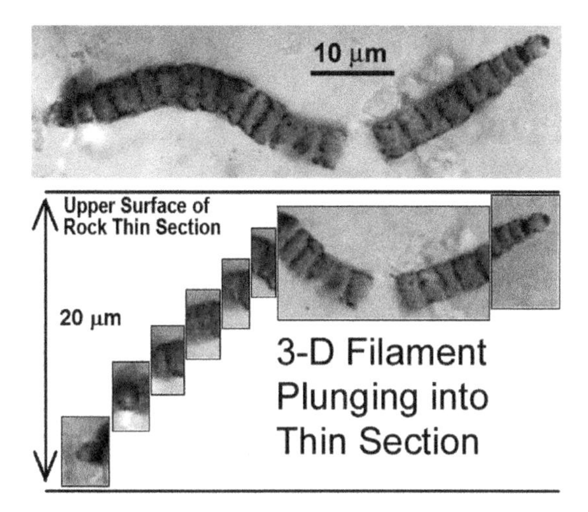

FIGURE 9.3 Precambrian cyanobacterial filament (800-million-year-old Bitter Springs Fm., Australia).

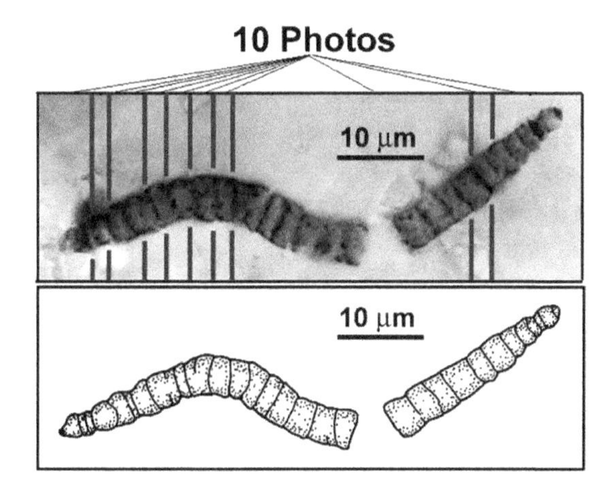

FIGURE 9.4 Photomontage of 10 photos of plunging Bitter Springs filament.

Though these were the best (in fact the only) effective techniques available, Bill was never fully satisfied because such artificial flattening results in distortion of the preserved cellular morphology (an undesirable degree of "nature faking"), particularly for pasted-together photomontages of sharply plunging fossil filaments. And because of the very thin (0.3-micron-thick) in-focus plane acquired at such high optical magnification, this photo-pasting technique provided useful composite images only of the thin in-focus central plane of such specimens, not of their entire cylindrical form. In addition, the technique was not applicable to spheroidal fossil unicells, the other common component of such ancient assemblages, for which publishable images could be obtained only of a single thin plane – usually of their equatorial plane in order to document maximum cell size. And the technique was even worse for depicting the colonies in which many such unicellular microbes typically occur, since the organization of their commonly numerous overlying and underlying cells could not be shown in a single composite image.

These deficiencies niggled at Bill for years and years. He experimented with modifications, additions, and wholly different techniques, but none resulted in improvement. Then, some 15 years ago, in the Biology literature, he ran across the newly devised technique of confocal laser scanning microscopy (CLSM), which after infusing living cells with laser-responsive fluorescent antibodies designed to bind to intracellular structures, could be used to image in three dimensions their internal membranes and organelles – their intracellular architecture. This technique, developed in 1985 by cell biologists at Cambridge University in England, produces high-resolution results for modern cells. But Bill, of course, wanted to image whole exceedingly tiny ancient fossils *in situ*, within a thin slice of rock, and the possibility of infusing a more or less hermetically sealed solid rock and its contained fossils with fluorescence-tagged antibodies was out of the question.

Fortunately, by this time, Bill had gained sufficient knowledge of the chemistry of kerogen, the coaly carbonaceous matter of which ancient microscopic fossils are composed, that he imagined that if the dominating polycyclic aromatic ring structures of the fossil kerogen were irradiated with an appropriate wavelength of laser light, they too would fluoresce and produce the images he sought. He gave it a try. His notion was correct (Figures 9.5 through 9.7). And one of his graduate students, Ahbi Tripathi (an engineer then on leave from NASA's Johnson Space Center and now a prime scientist at Elon Musk's SpaceX Corporation in Hawthorne, California), used this new idea to produce a fine Ph.D. thesis.

Bill's colleagues, students, and he have now used CLSM on many projects, including that establishing the cylindrical, cellular three-dimensional morphology of members of the 3,465-million-year-old microfossil assemblage of the Apex chert, which, although not nearly as well preserved as other billions-of-years-younger Precambrian fossils (Figures 9.5 through 9.7), are among the oldest evidence of life now known (Figure 9.8).

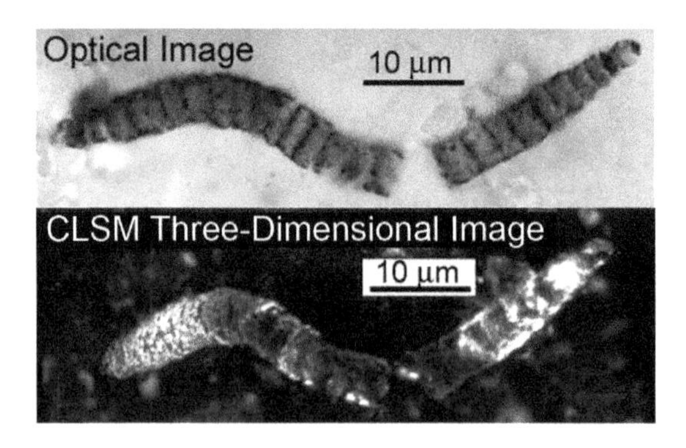

FIGURE 9.5 Optical and 3-D CLSM images of plunging Bitter Springs filament.

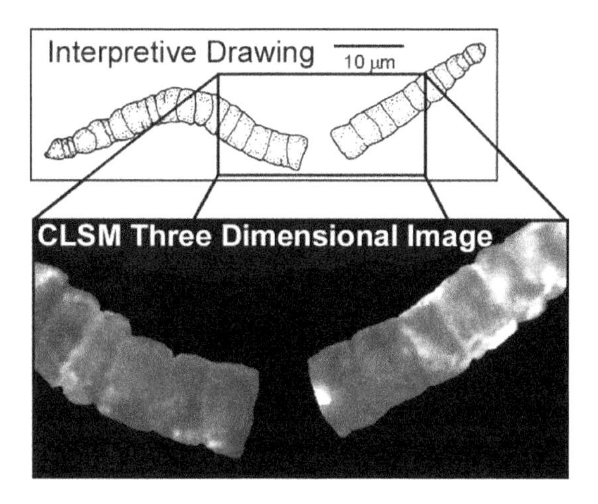

FIGURE 9.6 High-resolution CLSM image of central part of plunging Bitter Springs filament.

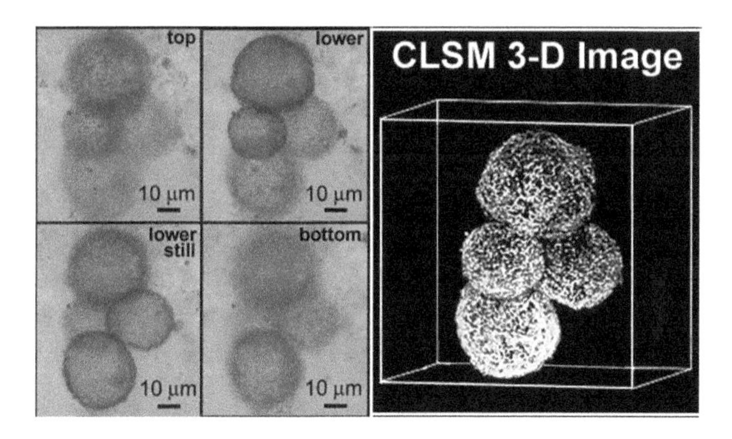

FIGURE 9.7 Optical and 3-D CLSM images of a four-celled colony (750-million-year-old Chichkan Fm., Kazakhstan).

9.5 RAMAN SPECTROSCOPY

Well before this, Bill had wondered about another big problem that had stuck in his craw since graduate-school days, namely, "How does one know that ancient microscopic fossils are actually composed of carbonaceous organic matter?" Visually, the evidence all fit: their brownish color and the texture and the shape of their cells was what should be expected, and they certainly look like (variably degraded) modern microbes. But that was about it – and such notions seemed to him to be incomplete "by-guess-or-by-golly" assessments of their composition, reasonable guesses unsupported by direct *in situ* chemical evidence.

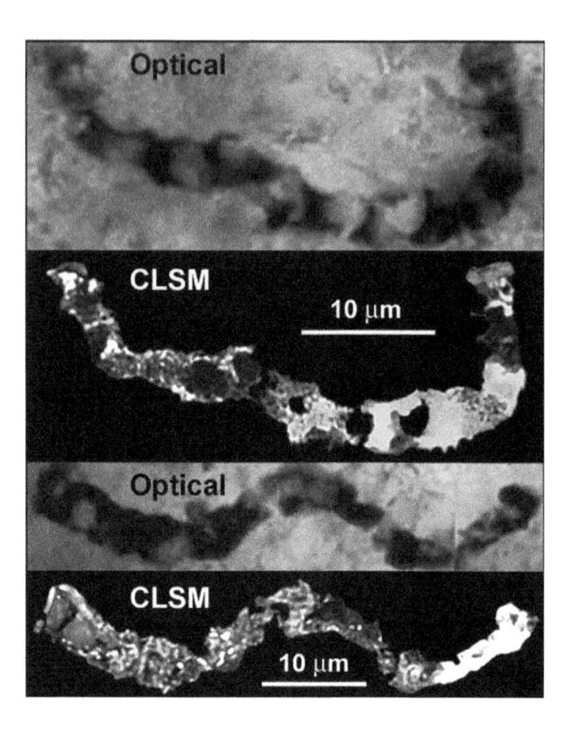

FIGURE 9.8 Optical and CLSM images of 3,465-million-year-old Apex chert filaments.

As a graduate student, he asked Professor Barghoorn about this problem. Barghoorn's response was the best then available: "In the coal industry they do an ash analyses," a process the two of them then repeated by dissolving a small sample of the mid-Precambrian Gunflint chert in mineral acid; drying and weighing the residuum; heating the residue in a crucible over a Bunsen burner; and then weighing the crucible a second time, the logic being that "the difference in weight shows that the carbonaceous matter [presumably that composing the fossils] has been volatilized, burned off."

OK ... Bill now knows that as indirect as this analysis may have been, Barghoorn was (mostly) correct. But at the time, this approach struck him as an "argument by analogy" – particularly the assumption that the chemistry of coal and its compression-preserved plant fossils was necessarily the same as that of the permineralized microbes of decidedly more ancient Precambrian cherts. This was a form of argument that, he had learned from his undergraduate Philosophy courses, was "notoriously suspect." Moreover, Harvard professor Barghoorn was using what Bill later came to refer to as "Science by Authoritative Assertion" – that is, don't question authority (in this instance, the coal geologists or a Harvard professor), it is always right! Neither of these arguments rings true (though at the time, of course, Bill said nothing of the sort). Such "ash analysis" provides no direct chemical–molecular evidence of the composition of a fossil embedded within a chunk of rock. And Bill also well knew that opaque minerals could mimic the color, a bit of the texture, and in rare instances even hints of the shape of authentic microscopic fossils.

A way to solve this conundrum ultimately emerged. In the late 1990s, Bill attended a scientific meeting at which, during a poster session, he came across a presentation by a group from the University of Alabama, Birmingham, that reported chemical analyses of organic matter in a carbonaceous chondrite (an organic-rich type of meteorite) carried out by Raman spectroscopy. Although the carbonaceous matter was in individual tiny bits and pieces, not enclosed in solid rock as are Precambrian microscopic fossils – and although Bill knew nothing at all about Raman spectroscopy – he nevertheless thought, "this has promise – perhaps I can use this technique to address my long-standing and festering problem."

Bill got in touch with the Birmingham group, prepared test specimens, and then flew off to Birmingham repeatedly, two or three times a year, as they perfected the use of the technique for permineralized rock-embedded specimens. As it turned out, Raman spectroscopy worked like a charm and is now a part of the "standard toolkit" for Precambrian paleobiologic studies worldwide.

Bill then got another break. Funding for the Birmingham Raman expert, Dr. Anatoily Kudryatsev, had expired, and he was looking for job. At about the same time, the one person Bill had hired to help him at CSEOL encountered difficulties with the UCLA administration. He departed, and Bill used the funds then available to pay half (due to well-merited salary increases, now down to 30%) of Kudryatsev's salary. Together, Bill and Anatoliy designed a laboratory in which they installed a new Raman instrument having all the "bells and whistles," which has been periodically upgraded ever since. Over the ensuing years, the two of them together have made "beautiful music" – among their many collaborative projects documenting *in situ* the carbonaceous (kerogenous) composition of innumerable ancient microscopic fossils, including those of the 3,465-million-year-old Apex chert. Here again, one of Bill's Ph.D. students, Andy Czaja greatly profited from this work; after a post-doctoral stint, he joined the faculty at the University of Cincinnati where he has established a Raman-CLSM lab much like that at UCLA and is now a tenured professor.

Raman is a highly versatile technique, excellent for documenting the molecular-structural chemistry of minerals (and for this reason is rather commonly available in university Geology departments). And because Raman can be used to analyze the chemistry of liquids and gases as well as solids, such instruments are also fairly common in Chemistry departments. From studies of literally scores of permineralized fossil assemblages, both of Precambrian and younger geologic age, Bill and Anatoliy discovered that the coal-like kerogen of which they are composed records a detailed record of their degree of geochemical alteration as geological pressure-cooking drives the organic matter toward its ultimate endpoint, the crystalline mineral graphite. From this, in 2005, they devised a quantitative measure of such alteration, the Raman Index of Preservation (RIP, a memorable moniker that Bill playfully coined from "Rest in Peace"), which documents the fidelity of preservation of the organic components and, thus, of the fossils themselves. This metric has been a great help to those studying such organic-walled fossils – a fast and easy way to determine how well they are preserved – as well as to the petroleum industry in its search for new promising sources of oil and natural gas (both of which hinge on the "geochemical maturity" of the preserved organics from which the oil and gas are derived).

Moreover, Raman, like CLSM, is a confocal technique – meaning that it blocks out signals from all but one thin plane of the material analyzed. And also, like CLSM, it can be used to scan across a given area of the fossil-containing plane, measuring hundreds of individual point spectra that together provide a complete two-dimensional chemical map of a thin in-focus slice of a fossil specimen and its associated minerals.

Given this, it occurred to Bill that, again like CLSM, it might be possible to vertically stack such two-dimensional chemical maps to prepare complete three-dimensional chemical images of such fossils. As he and Anatoliy got to work on the project, they discovered that it was technically more difficult than Bill had imagined, primarily because the individual thin-plane chemical maps, acquired sequentially at 0.2 µm at depths throughout a fossil, had to be stacked manually (rather than automatically as is the case for CLSM). For a typical 5-µm-thick Precambrian microscopic fossil, this required the 25 individual thin slices to be stacked into a perfect pile in which even the tiniest offset from one slice to the next would destroy the final image.

To defeat this problem, Bill contacted Jerry W. Gaskill, a long-time friend from Bill's Oberlin College days and the mathematician who had solved the far more complex problem of stacking in three dimensions the X-ray images used to produce cancer-detecting mammograms (a marvelous bit of wizardry that Jerry never patented, much to the consternation of his wife). Jerry told Bill, "That stacking business has been a long-standing difficult problem, but vertical stacking is pretty easy." Jerry then sent Bill a long letter detailing step-by-step how to solve the problem and also referred Bill to a website of the US Public Health Service, which included a free-of-charge computer program to address the matter.

Thanks to Jerry, this solved the problem, and Bill and Anatoliy then used this technique first to document in three dimensions the cell walls, carbonaceous composition, and chert-infilled cell cavities (cell lumina) at one end of a nicely preserved fossil filament from the 800-million-year-old Bitter Springs Formation of central Australia (Figure 9.9). Recognizing that this would be an important breakthrough in his science, Bill invited Jerry to be a co-author of the resulting article, but Jerry demurred, opining that "Oh no, I haven't done enough!" (much as Bill had told Professor Barghoorn years before, when it came to co-authorship of the 1965

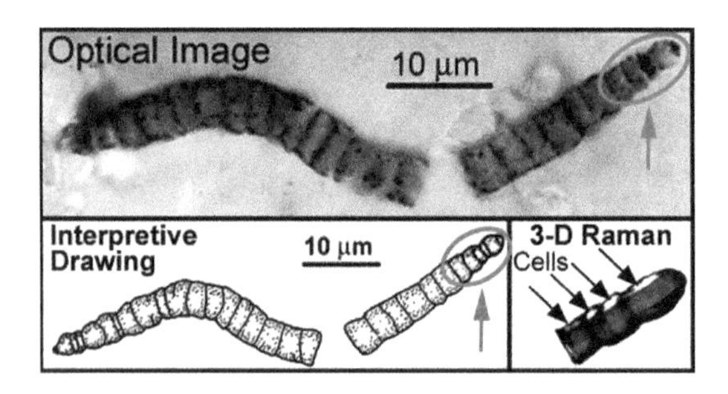

FIGURE 9.9 Three-dimensional Raman image of a Bitter Springs Fm. filament tilted to show the lumina of its end cells.

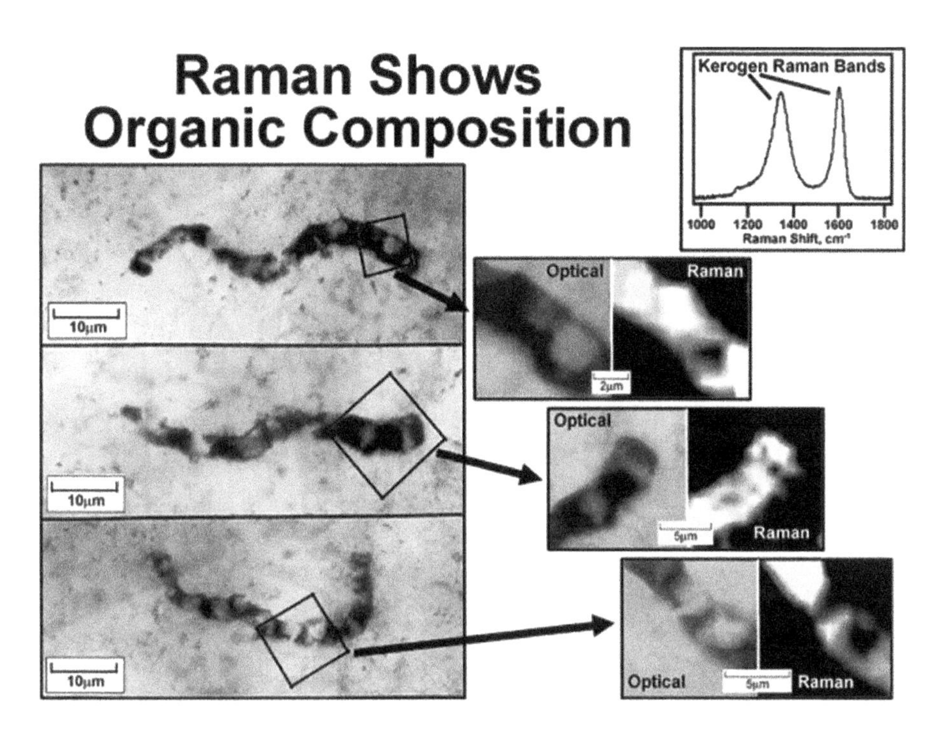

FIGURE 9.10 Optical and 2-D Raman images and a Raman spectrum showing the organic (kerogen) composition of Apex chert filaments.

Gunflint fossils breakthrough paper and, perhaps, an avoidance of presumed undeserved credit characteristic of many Bill's Oberlin cohorts).

In any case, in 2006, Bill and Anatoliy received a prize for this innovative work from the HORIBA Jobin Yvon Company, a manufacturer of Raman spectrometers, an advance now incorporated into the latest generation of such instruments. And, yet again like CLSM, they have used Raman on many projects, including 2-D Raman to establish the carbonaceous coal-like composition of the 3,465-million-year-old Apex chert fossils (Figure 9.10) and 3-D Raman to show their cylindrical form and three-dimensional cells (Figure 9.11).

9.6 FLUORESCENCE SPECTROSCOPY

One of the beauties of having a Raman instrument (at least like that at Bill's lab) is that it can also record the fluorescence spectrum of permineralized rock-embedded fossils. For particular scientific questions, this can be important, as is illustrated by the following example.

For quite some time (stemming from a paper by John R. Nursall that appeared in 1959), Bill has been interested in quantifying the oxygen content of Earth's atmosphere at the Precambrian-ending "Cambrian Explosion of Life," relatively large amounts of oxygen being needed by the diverse many-celled animals of this newly evolved biota. In terms of ecology and the carbon cycle, this makes good sense.

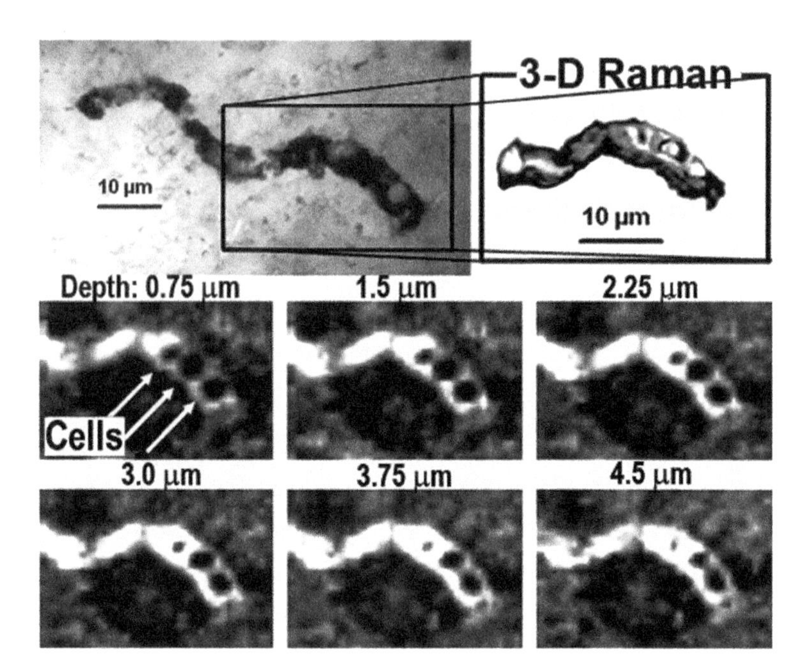

FIGURE 9.11 Three-dimensional Raman images showing the cylindrical shape and cell lumina of an Apex chert filament.

The light-powered chemistry of oxygen-generating photosynthesis, which since its origin billions of years ago has been the central driver of Earth's carbon cycle, combines the carbon and oxygen of carbon dioxide (CO_2) with hydrogen split away from water (H_2O) to form glucose sugar ($C_6H_{12}O_6$) and, as a by-product, produces oxygen gas (O_2) – in shorthand, $CO_2 + H_2O$ + light energy → "CH_2O"+ O_2.

But aerobic respiration (aerobiosis), the "breathing" of animals – the same process that occurs during the aerobic fungal and bacterial decay of dead carcasses – is the exact opposite of oxygen-producing photosynthesis: organic matter ("CH_2O") + and oxygen are consumed and carbon dioxide and water are given off, exhaled (just as you are doing as you read this passage). What this back-and-forth, in-out equality between photosynthesis and aerobiosis means is that if all the organic matter of a dead organism decays away – as almost always happens (the reason that our surroundings are not covered hip-deep in undecayed corpses of dead plants and animals) – there could have been no build-up of atmospheric oxygen.

In fact, the only way for oxygen to have built up over geological time is for dead organisms or their bits and pieces to have become buried in sediments where there is no oxygen and no oxygen-using decayers to use them as fodder.

Here's a way to understand this seemingly remarkable situation. There are two stable isotopes of carbon, ^{12}C and ^{13}C. Think of them like flavors of ice cream (their isotopes being "flavors" of carbon) – for example, plain vanilla ice cream and vanilla chocolate-chip. A spoonful of the plain vanilla will weigh slightly less than a spoonful of the chocolate-chip, which contains extra, heavier, bits of chocolate. In the same way, the "vanilla" isotope of carbon, ^{12}C, having an atomic mass of 12, is just a bit lighter than the "chocolate-chip" variety, ^{13}C, which is heavier, having an atomic mass of 13, because it has an extra particle in its atomic nucleus. Of the two carbon isotopes, ^{12}C is by far the more abundant, constituting 98.9% of the total (similar to vanilla ice cream, which is a lot easier to find at a store than its vanilla chocolate-chip close relative). Many elements have such atomic weight-defined isotopic siblings. Oxygen, for example, has three stable isotopes, ^{16}O, ^{17}O, and ^{18}O, of which the lightest isotope, ^{16}O, is overwhelmingly the most abundant, making up 99.8% of the mix.

Thus, because ^{12}C and ^{16}O are by far the most abundant isotopic "flavors," for all practical purposes, carbon has an atomic weight of 12 and oxygen, 16. It therefore follows from the carbon recycling back-and-forth balance between oxygen-producing photosynthesis ($CO_2 + H_2O$ + light energy → "CH_2O"+ O_2) and oxygen-consuming aerobic decay ("CH_2O"+ $O_2 \rightarrow CO_2 + H_2O$) that for every 12 grams of organic carbon that is buried in sediments, and therefore not recycled, 32 grams (2 × 16) of oxygen gas, O_2, accumulates in the environment. For virtually all of the Precambrian, the buried carcasses available for carbon recycling were microscopic organisms. But with the rise of animals, a new concentrated source of buried organic carbon came to the fore – not the bodies of the organisms themselves, which typically decayed away at the oxygen-containing sediment surface – but their "poop," their excreted fecal pellets!

Animals move. For this, they need energy. So they eat – a lot – but a sizeable portion of the food they take in is not totally digested. Because of this, their sediment-deposited fecal-pellet excretions, commonly enshrouded in gelatinous goo, vastly upped the rate of carbon burial and, thus, increased the rate of oxygen build-up. The atmosphere changed and animal life thrived, a rather remarkable environmental–biological, abiotic–biotic interaction. (But that, of course, is the nature of Darwinian evolution – life responding and adapting to its ever-changing environment, regardless of the cause of such changes.)

As chance would have it, coincident with the rise of animal life there was a huge influx of submarine volcano-derived deep-sea phosphate-bearing waters into shallow-water settings. This global event resulted in the embryos and sometimes even the body parts of such early-evolved animals – as well as the myriad microbes coexisting in their local environs – being permineralized by phosphate, specifically the mineral apatite. Apatite is a fairly complex mineral having the general formula $Ca_5(PO_4)_3(F,Cl,OH)$, its most commonly occurring forms being fluorapatite [$Ca_5(PO_4)_3F$], chlorapatite [$Ca_5(PO_4)_3Cl$], and hydroxyapatite [$Ca_5(PO_4)_3OH$], the last form being the mineral that makes up animal bones (including yours and mine). Among mineralogists, apatite is famous for its propensity to allow the substitution of other elements for those in its standard make-up. And of these various substituting elements are the so-called "rare earth elements" (REEs) such as samarium (Sm), europium (Eu), neodymium (Nd), and 14 others of which most of us have rarely heard.

Of particular interest for unraveling oxygen levels at the Precambrian-to-Cambrian transition is the REE samarium, which substitutes for calcium (Ca) atoms in the crystal lattice of apatite. There are two different types of calcium sites (known as Ca-1 and Ca-2) peppered throughout the apatite lattice, the sites differentiated from each other by the types and spacing of the atoms immediately surrounding each site. Samarium substitutes for calcium at both types of sites, but because of the chemical reactions that accompany such substitutions, the Sm-replacement of Ca-1 occurs in the absence of oxygen, and that of Ca-2 only if oxygen is present. And unlike unsubstituted apatite, the samarium-including form fluoresces under appropriate wavelengths of laser light. Not only does it light up in CLSM, but each site also has a distinctive fluorescence spectrum. And because the spectra of both sites can be acquired simultaneously, their peak heights (reflecting the relative amount of Sm-substitution) can be readily compared and quantitatively measured.

Great! Taken together, all this suggested to Bill that fluorescence spectroscopy of fossil-permineralizing apatite in specimens spanning the latest Precambrian into the earliest Cambrian – and, thus, the Cambrian "Explosion of Animal Life" – might provide a means to produce the first quantitative oxygen paleobarometer known to science, which could tell us, by direct measurement, the extent to which a rise in environmental oxygen may have been associated with this major evolutionary event. At present (January 2018), one of Bill's graduate students, Amanda Kathryn Garcia, is following this up. They are optimistic – things look good. [Added note: She did the job, proved the case, and is now Doctor Amanda K. Garcia. *Hooray for her!*]

9.7 ANALYTICAL TECHNIQUES AND THE SEARCH FOR PAST LIFE ON MARS

Among the truly great benefits of combining traditional high-resolution optical microscopy – the basic tool of the science – together with the relatively new techniques of CLSM and Raman and fluorescence spectroscopy is that the same individual specimen can be analyzed by all four techniques. Used in sequence, this is a great boon to sorting out *bona fide* ancient fossils from inorganic purely mineralic "look-alikes."

Each of these techniques provides benefits. In particular, optical microscopy can be used to locate what look to be good fossils and record their color, texture, and cellular and organismal morphology – but it cannot document their chemical composition. CLSM can then document at even higher spatial resolution their detailed three-dimensional cellularity and morphology and show them to fluoresce under the CLSM laser beam, a strong indicator of their polycyclic aromatic ring structure–dominated carbonaceous (kerogenous) composition. Raman, then, can confirm their biological origin by providing definitive evidence of their molecular structural composition and geochemical fidelity of preservation as well as document the mineralogy of their embedding matrix – and, if need be, map the specimens chemically in both two and three dimensions. For special problems – construction of a quantitative apatite-based oxygen paleobarometer, for example – fluorescence spectroscopy is the obvious tool of choice. Importantly, all of these techniques mesh together hand-in-glove, mutually reinforcing evidence derived from the others.

There is another great benefit of the combined use of these techniques, namely that all of them – optical microscopy, CLSM, and Raman and fluoresce spectroscopy – are non-intrusive and non-destructive. In other words, for the quartz- (or apatite-, or gypsum-, or calcite-) permineralized fossils now analyzed from a very wide range of geological samples and ages, the techniques penetrate deeply into the rock slice studied (to a depth of 150 microns, the combined thickness of five standard petrographic thin sections) and, while properly applied, they in no way alter the rock or its contained fossilized microbes.

This benefit is by no means trivial. When the rocks to be cached by NASA's Mars2020 Mission are finally brought to Earth (in 2030, 2040, who knows?) they will be the most precious materials on the planet. The total cache will be small – perhaps only 100 or 200 grams of rock – but each single gram will have cost some *50 to 70 million dollars* to acquire! Heavens! – no pun intended – they will be more valuable than all the Queen's jewels and the all the glorious golden relicts of King Tut's tomb, combined! Science will face a conundrum. The rocks *must* be studied to determine whether they harbor evidence of ancient Mars life, the prime question to be addressed in this venture. But they also *must* be protected, not harmed in any way, so that scientists asking other questions can use the same rock and discover their answers, too. The tandem use of non-intrusive and non-destructive optical microscopy, CLSM, and Raman and fluoresce spectroscopy will meet this need.

9.8 SECONDARY ION MASS SPECTROMETRY

In 2000, one of Bill Schopf's students and he made yet other important technique-based contribution. From Bill's earliest days as a graduate student – stimulated largely by the pioneering 1960s work of Tom Hoering at the Carnegie Geophysical Laboratory in Washington, D.C. – he had realized that measurement of the carbon isotopes in bulk samples of kerogen extracted from ancient rocks was terrifically useful. Unfortunately, however, there were only a limited number of measurements. So, this was a problem that Bill structured the first Precambrian Paleobiology Research Group (PPRG) project to address. It did so by analyzing a large number of deposits that extended the isotope-based record of photosynthesis back through time to more than three billion years.

Yet despite this important new knowledge, to Bill it seemed insufficient, since such measurements provide only an *average* value of the carbonaceous matter in an ancient rock, material derived from a great mix of physiologically diverse biological sources typically dominated by photosynthetic CO_2-consuming carbon-fixers. Instead, what he wanted was measurement of the carbon isotope composition of single minute fossils, fine-scale detailed data that, in principle, could reveal the physiology of such microbes that could then be linked to *individual* morphology-defined fossil species.

In the mid-1990s, Bill got lucky. The then-new technique of secondary ion mass spectroscopy (SIMS) seemed to offer the solution he was looking for, and one of his department colleagues (Professor Kevin McKeegan) had recently acquired the hundreds-of-thousands-of-dollars instrument needed to do the work. Bill prepared a set of samples and put graduate student Chris House on the problem (a truly excellent student who, as an undergraduate at the University of California, San Diego, had worked in the lab of Bill's Origin-of-Life friend Stanley Miller, who had telephoned him repeatedly, urging him to accept Chris as a graduate student). Chris worked assiduously over many months, waiting his turn as the geochemists in charge of the SIMS instrument used it for zircon-based age-determining geochronology (their prime interest) as well as for non-UCLA users who paid for time on this then-rare new instrument. Finally, during a few overnight sessions on weekends when the instrument was not otherwise occupied, Bill's student got his chance. The data were sound and the project was completed. Chris House (who has since become a professor at Penn State University and a major player in Precambrian Paleobiological science) received a well-earned Ph.D. In 2000, the results were published.

More recently – and to Bill, very gratifyingly – yet another of his former students, Dr. Dorothy Oehler, has become a notable leader in the use of SIMS (thanks largely to her collaboration with Françoise Robert, a first-rate isotope geochemist and a friend of Bill's in Paris). After a stint in Australia and a long-time affiliation with Conoco Inc., Dorothy and her husband (the gifted novelist John Oehler, another of Bill's Ph.D. graduates) came to reside in Houston, Texas, where for the past decade or so, she has held positions both at NASA's Johnson Space Center and, at present, the nearby Lunar and Planetary Institute. Dorothy claims that she was Bill's very first doctoral product. Maybe so – he had three in the mill at that time, she, her husband John, and Bob Horodyski – and Bill honestly does not recall who among them was first to get out of the nest. The salient point is that Dorothy, an excellent student, is still plenty active in the game. All in the field cheer her on – with Bill serving as the lead cheer-leader, knowing full well that she is only four or five years younger than is he.

This fourth technique, SIMS, is not without drawbacks. Not only is the instrument extremely costly but its use also requires the target fossils to be exposed at the upper surface of a polished thin section, so that charged ions (e.g., $^{12}C^{-}$, $^{13}C^{-}$) sputtered off the specimen can be swept through the main mass spectrometer of the system, enabling their identification by their mass (i.e., atomic weight)-to-charge ratio and their relative abundances to be determined. This surface-exposure requirement limits the availability of analyzable specimens to only a smattering of fossils in a given thin slice of rock. Moreover, SIMS instruments are plagued by the notoriously poor resolution of the attached optical microscopes that are used to locate the selected targets, a significant problem for studies of exceedingly tiny Precambrian microbes. And finally, unlike CLSM or Raman or fluorescence spectroscopy, analyses by SIMS produce small pits (about 10μm across and a few microns deep) in the surface of the sample analyzed. Given this, it seems likely that SIMS analyses will not be permitted for rock samples from Mars (unless, of course, the analyses are judged to be crucially important, at which time NASA can be predicted, as in the Apollo 11 and 12 studies, to issue an "Official Waiver").

Despite these drawbacks, SIMS has played a decisively important role in studies of Earth's earliest fossils. In Chris House's doctoral research, it was used to study 15 chert-permineralized fossils of the Precambrian Bitter Springs Formation, the carbon isotope signals of which could then be compared with 32 analyses amassed by the PPRG of the carbonaceous matter extracted from kilogram-sized bulk samples. The ranges and averages of the two sets of samples were very nearly identical, establishing the veracity of this newly introduced technique.

Over subsequent years, Bill has used SIMS repeatedly, largely in collaboration with Professor John Valley and his colleagues of the WiscSIMS Lab at the University of Wisconsin-Madison, analyzing both carbon and sulfur. Recently, in 2015, their team used such studies to back Bill's finding of 2,300- and 1,800-million-year-old examples of a sub-seafloor sulfur-cycling ecosystem previously unknown in the geological record and their relevance to understanding the "null hypothesis" required of Darwinian evolution. Then, more recently in 2017 and 2018, SIMS data were crucial to two additional breakthrough finds, both in especially ancient Archean-age (greater than 2,500-million-year-old) deposits: first, SIMS measurements of the sulfur-signatures of 3,400-million-year-old mudflat-inhabiting microbial communities showing that Earth's primordial atmosphere lacked oxygen; and second, SIMS carbon-measurements documenting the metabolism of five species of microbes in the 3,465-million-year-old Apex chert, for the first time establishing the presence in this oldest-known diverse fossil assemblage of cellularly preserved photosynthetic microbes and early-evolved methane producers and consumers – the same early-evolved microbes that knowledgeable microbiologists had imagined to have played a major role in Earth's early ecosystem. Without SIMS, there would have been no way to seal the deal.

Over his career Bill also "pioneered" (i.e., was the first to apply) various other then-new techniques to studies of Precambrian microscopic fossils – among these transmission electron microcopy (1965), electron probe microanalysis (1971), and atomic force microscopy (2005) – each providing excellent data for particular types of problems. But, coupled with the photomicrographic montage technique he developed as a graduate student, probably the most useful techniques his students and he have introduced to the field are CLSM, Raman, and SIMS, all of which are the preferred analytical techniques currently in use worldwide.

9.9 THE SPREAD (OR LACK THEREOF) OF KNOWLEDGE IN SCIENCE

Over Bill's career – and given his values – he became increasingly dismayed by the lag-time (not uncommonly as many as 15–20 years) between the introduction of new instrument-dependent techniques and their widespread adoption, especially in developing countries. To address this problem, in 2008 he convened at UCLA's Center for the Study of Evolution and the Origin of Life (CSEOL), a week-long "World Summit on Ancient Microscopic Fossils" attended by 31 leading scientists from 12 countries (Australia, Brazil, Canada, China, England, France, India, Japan, Russia, South Africa, Sweden, and the United States). The results have been gratifying – CLSM and Raman instruments are now available for such work at the Birbal Sahni Institute of Palaeobotany in Lucknow, India; and are currently in the process of being purchased both by the Nanjing Institute of Geology and Palaeontology, China, and the Geological Institute in Moscow, Russia. In addition, spurred by the summit and Bill's later two-week lecture trip to São Paulo, his colleagues in Brazil are now active users of CLSM.

This international meeting at UCLA had its desired effect – CLSM and Raman for general use (and fluoresce spectroscopy and SIMS for more specialized applications) are spreading rapidly. But the science is now only a decade distant from the 2008 meeting, and this is just the beginning. All of these techniques are bound to spread, hopefully in concert, worldwide. In fact, there will be no choice. Knowledgeable scientists will be aware that the techniques exist, and if the telling data they can potentially provide are not included in the studies carried out, the incomplete results submitted for publication will be rejected out-of-hand. Difficult as this may be to those lacking access to the required instruments, it is the way of science. And as Bill Schopf has repeatedly suggested to others, the problem can easily be defeated by interregional and international collaboration which, by spurring interaction and cooperation, will be good for the science and the furtherance of knowledge.

9.10 ANALYTICAL TECHNIQUES: PERSEVERANCE AND SOURCE

As you may have gathered, Bill Schopf is a "try to solve a problem, try to figure out an answer" type, not an "instrument guy." The difference between these two approaches was impressed on him during his graduate-school days. At that time, transmission electron microcopy (TEM) was brand new – though Harvard, of course, already had a couple of these new instruments (initially housed in the solidly emplaced basement of the Biology Building because stable shock-absorbing platforms had not yet been perfected). Bill ventured over to that lab in hope of getting the Harvard experts to help him understand what TEM could provide for his Precambrian microfossil studies. They were welcoming, kind, and delighted. They did some work and Bill got out two scientific papers. Good stuff for a mere graduate student! But what he also learned – a lesson that has stuck with him all these years – is that the instrument operators had only a very limited idea of the scientific problems to which TEM could be applied. As accomplished as they were, they were technicians, instrumentalists, not see-through-the-problem scientists.

Years later, Bill discovered this same syndrome among his instrument-building fiends at NASA-JPL. They had recently developed a planet-orbiting infrared system having ten-meter Mars-surface spatial resolution. The problem was that the questions posed by JPL scientists needed one-meter resolution, not ten, to examine the distribution of craters formed by impacts of bits and pieces of impacting meteors. Instead of setting about to increase the resolution of their system by the order of magnitude the scientists needed, the instrumentalists asked Bill for what problems their less-than-optimum system could be used.

It may not be welcome to note the following, but both of these episodes – that with Bill's JPL colleagues and his earlier Harvard TEM experience – seemed to him to be inappropriately reversed "cart-before-the horse" approaches to science. For knowledge to progress, it is the questions about the real world that must drive the work, not the fitting of the questions to the availability of some instrument, however new and fancy. To Bill's mind, this is a fundamental difference between question-asking, fact-seeking scientists and instrument-designing or equipment-overseeing instrumentalists. For that reason, and despite Bill's introduction to the field of a number of new-to-the-science analytical techniques, his work has focused on questions and facts rather than on the latest available equipment.

Two meaningful notions are suggested by all of this:

1. Over time, ideas that occur early during a young scientist's career can ultimately come to fruition given continued perseverance – illustrated by Bill's long-term search for methods to analyze accurately and *in situ* the three-dimensional morphology (CLSM) and both the molecular (Raman) and isotopic compositions (SIMS) of individual microscopic fossils. To him this suggests that, as apocryphal as it might seem, there is a real advantage to being a less-than-knowledgeable novice – and, as its corollary, that many academics do a disservice to their imaginative younger inexperienced students by permitting them to "sell themselves short."
2. Useful techniques can be "borrowed" from seemingly unrelated fields – for Bill illustrated by the applications to paleobiologic problems of CLSM, designed initially for neobiological studies of the interiors of living cells, not paleobiological studies of billions-of-years-old fossils; and also of Raman and SIMS, techniques originally designed to be used in geochemistry rather than analyses of tiny individual fossilized microbes – applications that their practitioners had presumably not previously imagined.

In short, this question-first, instrument-second approach has been a key to the success of Bill Schopf's science, perhaps coupled with what others seem to regard as his ability to "think out of the box," to ask new questions, and to be aware of advances in ancillary fields – and to then apply the results, as appropriate, in ways that others had not yet attempted.

9.11 HOW HAS 50 YEARS OF THIS SCIENCE "ARTICULATED" THE NEW PARADIGM?

In the Kuhnian sense specified in *The Structure of Scientific Revolutions*, the paradigm-shifting solution to Darwin's dilemma – from an "unknown" to a known record of Precambrian life, the seven-fold extension of life's long history from 500 to 3,500 million years ago – began in the 1950s with Stanley Tyler's discovery of the mid-Precambrian permineralized microbiota of the Gunflint chert in southern Canada, Boris Timofeev's seminal finds of compressed microfossils in Precambrian shales of Siberia, and Martin Glaessner's convincing documentation of the very latest Precambrian age of the imprints of soft-bodied metazoans that comprise the Ediacaran Fauna of South Australia. But the most influential paradigm-changing breakthroughs date from 1965 and the publication by Elso Barghoorn and Stanley Tyler of descriptions of the Gunflint microfossils, the almost immediate validation of the fossils by Preston Cloud, and the report by Barghoorn and Schopf a few months later of the discovery of a second but decidedly better-preserved microbial biota in cherts of the late Precambrian Bitter Springs Formation of central Australia (the subject of Bill Schopf's doctoral thesis, published in 1968). Considered as a whole, this series of contributions showed finally, once and for all, that evidence of Precambrian life is not only detectable but can be readily deciphered.

Using Kuhnian phraseology, all studies subsequent to these seminal 1965 contributions fall into the category of "articulating the paradigm," workers conducting what Tom Kuhn referred to as "normal science." Whether regarded as normal, standard, or even commonplace and run-of-the-mill, such studies are the very essence of the scientific enterprise. As Kuhn clearly demonstrated, these subsequent studies extend the paradigm, fill in the gaps, and evidence with find after find after find that the paradigm shift resulted in a more complete description and understanding of reality. That is the way that science works.

Below are recorded some of the more notable contributions of the past half-century to the articulation of this new paradigm, findings that seem likely to stand the test of time. But it is appropriate to remember that the science of Precambrian Paleobiology is still a mere 50 years young. Its findings are firmly established and universally accepted – the track record of the science shows that it is no "flash in the pan," no great all-embarrassing mistake that will ever need to be retracted. Nevertheless, a scant 50 years of study of a record of life that extends to at least 3,500 million years ago, seven times longer than all of later evolutionary history put together, can be expected to have revealed only the tip of a huge underlying iceberg. Understanding of the interrelated Precambrian history of life and its environment is still relatively "new knowledge." Much remains to be learned.

Pause for a moment to think about what this means and recall the adage of mathematician and philosopher Alfred North Whitehead: "Almost all new ideas have a certain aspect of foolishness when they are first produced." Note also that the listing below of contributions that appear to be of lasting value is meant to look well into the future, perhaps decades from now, and about which none of us today can have firm understanding, our descendants by then routinely doing science that to us would seem nearly indistinguishable from magic. If at first glance this seems futuristic nonsense, imagine your great-grandparents and how they would have regarded television, superhighways, driverless automobiles, computers, the internet, cell phones. And if the Whitehead quote does not fit your fancy, recall Yogi Berra's malapropism that "It's tough to make predictions ... especially about the future."

In short, though virtually all of the advances listed below are now widely accepted, some are more firmly established than others. The following is a synopsis of ten items that are intended to encapsulate the current status of the field and the major paradigm-articulating advances in its yet young history:

- At its inception an area of concerted interest to only a handful of scientists worldwide, Precambrian Paleobiology is now broadly interdisciplinary and international, several hundred workers in many countries making discoveries, sharing expertise, and working together to further the science.
- Over the now more than 50 years following the final paradigm-changing finds of 1965, understanding of microbially produced stromatolites has increased exponentially – now known from literally hundreds of deposits, both modern and Precambrian – spurred initially by a splendid monograph published in 1976. Analyses of molecular biomarkers have advanced by

leaps and bounds, as illustrated, for example, by the 1988 discovery of the hydrocarbon biomarker gammacerane, likely the hydrogenated derivative of tetrahymenol produced by ciliate protozoans, in a Precambrian geologic unit of the Grand Canyon, Arizona and a discovery that meshes with the earlier (1973) finding of the oldest known fossilized protozoans in the same deposit. And several hundred microfossil-bearing Precambrian deposits and well over 1,000 new biological species have been discovered, the science guided by the ground-breaking international interdisciplinary PPRG team, which, by its work from 1977 to 1992, defined and focused the science and played a major role in advancing the field to its current status.

- Pivotal to this progress has been the introduction of new analytical techniques now used widely for the study both of rock-embedded and compression-preserved Precambrian microorganisms, including TEM and AFM but most notably CLSM, Raman spectroscopy, fluorescent spectroscopy, and SIMS, techniques that hold promise in the search for evidence of ancient life in rocks from other planets.
- Spurred by the availability of such new techniques, the field has tackled such fundamental problems as documentation *in situ* of the organic (kerogenous) composition and metabolism-evidencing isotopic compositions of ancient microbial fossils; their mode of preservation, the mineralogy of their associated rock matrix, the geochemical maturity of their kerogenous components, and their fidelity of preservation; their organismal form and three-dimensional cellular anatomy; and – of critical importance – their biogenicity and indigenousness to and syngenicity with the formation of the rock in which they occur.
- The initial find of permineralized Precambrian microscopic fossils in the Gunflint chert of southern Canada and its confirming find in central Australia have been extended to include discoveries of similarly or appreciably more ancient fossils in Australia, China, Europe, India, Russia, South Africa, and South America, including an especially ancient fossil assemblage from northwestern Australia that extended the known antiquity of life to nearly 3,500 million years ago (supported by other microfossils and stromatolites of similar age).
- The findings of Precambrian Paleobiology have contributed greatly to the field of Evolutionary Biology by providing new insight into (1) the extinction of large-celled phytoplankton followed by their rebound and the rise of animals at the "Cambrian Explosion of Life" about 550 million years ago; (2) the fossil record–evidenced diversification of nucleated (eukaryotic) phytoplankton, multicellular algae, and protozoans, spurred by the advent of eukaryotic sexuality about 1,000 million years ago; (3) the exceedingly slow (hypobradytelic) evolution of cyanobacteria over an enormous segment of Precambrian Earth history; and (4) the metabolic, biosynthetic, and organismal responses to the mid-Precambrian global increase of atmospheric oxygen (the Great Oxidation Event), which, together with the

earlier origin of photosynthesis and the later advent of eukaryotic sexual reproduction and its products – large multicellular organisms and the gradual development of an increased ability to interact with and manipulate the environment ("intelligence") – is one of the few truly major events in the history of life.

- The field has been replete with unexpected new findings, such as the discovery in mid-Precambrian sediments of a previously unknown sub-seafloor sulfur-cycling ecosystem and its use to provide the first promising evidence (after 150 years) of the "null hypothesis" required of Darwinian evolution. Other such "firsts" relevant to life's even earlier stages include discovery of (1) evidence encoded in reconstructed ancestral enzymes indicating that over geological time, Earth's surface temperature cooled from about 70°C (158°F) some 3,000 million years ago to its present 14°C (57°F); (2) a 3,400-million-year-old mudflat-inhabiting microbial consortium of sulfate-reducing bacteria that produced H_2S consumed by coexisting primitive (non-O_2-producing) photosynthetic bacteria, showing that the early Earth lacked all but traces of atmospheric oxygen; (3) 3,465-million-year-old methane-producing Archaea that generated CH_4 consumed by intimately associated methane-consuming microbes; and (4) diverse lines of evidence establishing that primordial life originated earlier and evolved farther and faster than had previously been imagined.
- By combining traditional Paleontology with Microbiology, Isotopic and Biomarker Geochemistry, Paleoecology, Atmospheric Evolution, Phylogenetics and Genomic Molecular Biology, and Evolutionary Biology, Precambrian Paleobiology has illustrated a new, effective way to unravel the interrelated history of life and Earth's environment. The success of this initially rather novel interdisciplinary approach can be expected to spur a new wave of advance in the present "Century of Genetic Engineering" (which also holds the promise of being a "Century of Evolution," of interdisciplinary progress toward understanding interrelated biotic–environmental history over geological time).
- The findings of Precambrian Paleobiology have set the stage and provided the basis for the new field of Astrobiology by establishing, for example, that (1) sedimentary gypsum, a rock type widespread on Mars and previously largely ignored by the paleontologists, is capable of harboring excellently preserved fossil microbes; (2) relevant to sample-return missions, a miniscule (smaller than postage-stamp) rock sample can contain firm evidence of diverse billion-year-old microbes; (3) even primitive biological systems are capable of evolving far and fast; and (4) like life on Earth, living systems may have originated and flourished on other planets within a few hundreds of millions of years after planetary formation.
- As a result of the example set by the interactive, interdisciplinary, international PPRG, this paradigm-changing science has provided the bases both for NASA's highly effective Astrobiology Institute (NAI) and the selection of rocks to be cached by NASA's Mars 2020 Mission for later in-depth study on Earth in the search for evidence of past life on Mars.

9.12 WHERE MIGHT THIS SCIENCE GO FROM HERE?

While none of us can claim to be an infallible sooth-sayer, able to accurately predict how this science will progress, some among us know its past and present, have a good feel for what now is known and what is not, and have educated notions about some of the pressing unsolved problems that remain to be tackled. The following are at the top of Bill Schopf's yet-to-be-accomplished "to do list."

9.12.1 Changes in Day-Length and UV-Flux Over Geologic Time

Since 1962 and the pioneering studies of the daily growth increments of fossil corals carried out by Cornell University paleontologist John Wells, it has been known that during the Silurian Period of geological time, some 420 million years ago, each year contained 400 days, not the current 365. Clearly, Earth's year-defining rotation has slowed over the geological past, the slowing a result of the frictional tidal sloshing of oceans against their margins as the Earth spins on its axis. Wells' implied estimates of the rate of slowing were soon backed by analyses of 2,500 BC manuscripts that record Chinese solar eclipses observed by sky-watchers from a rooftop in the Emperor's "Forbidden City," which evidence a decrease of Earth's spin rate by a few milliseconds per century.

Because the Moon is locked in Earth's gravitational field, as the Earth slows, the Moon gradually drifts away. In essence, this phenomenon of the Moon moving away as Earth's spin gradually slows (predicted by Isaac Newton's "Law of the Conservation of Angular Momentum") is simply the reverse of what an Olympic ice-dancer spectacularly demonstrates when she starts her spin slowly, with her arms far-outstretched, and then spins faster and faster as her arms move to clutch her body. The gradually increasing distance between Earth and Moon is now known to be about four centimeters per year, determined by the time traversed by laser beams traveling at the speed of light bounced off reflectance arrays that were placed on the Moon by the Apollo 11, 14, and 15 astronauts and the two 1970 and 1973 unmanned Soviet Lunokhod missions. And these measurements in turn provide the means to calculate quite precisely the imperceptible lengthening of Earth days.

Logically, then, all one would have to do to determine day-length over the geological past would be to extrapolate the current rate of slowing into deep geological time and calculate the ever-increasing number of days per year and the correspondingly fewer hours per day. This, however, presents a problem. As the time before the present recedes into the past – and the calculated days per year increase and day-length decreases – the distance between Earth and Moon must also decrease. Between and two and three billion years ago, during the mid-Precambrian, when the calculated day-length would have been much shorter than today, the Moon would have been so close to the Earth as to have reached the Roche limit, a distance of about 11,000 miles (approximately two-and-a-half times Earth's radius), and the tidal pull of the Earth would have equaled the gravitational attraction of the Moon to itself. The internal gravity that holds the Moon together would have been overwhelmed and the Moon would have fallen apart, torn asunder by Earth's tidal forces with chunks, bits, and pieces raining down across the Earth in a huge planet-wide meteor shower.

An event such as this may have happened to one or more moons that once orbited Saturn, a plausible source of at least some of Saturn's famous rings, but it obviously did not happen here. Earth's Moon still exists!

Why does day-length matter, and how can its real history be deciphered? Day-length is crucial to all photosynthetic (and therefore light-dependent) forms of life, not only to the more familiar land plants and seaweeds but to their evolutionary ancestors such as cyanobacteria and their precursors, anaerobic photosynthetic bacteria, the two types of early-evolved photosynthesizers that were the ecosystem-dependent "primary producers" throughout virtually all of the Precambrian. Moreover, the history of day-length encodes a record of the changing distance between the Moon and Earth. Yet extrapolation into the distant past of the current rate of Moon's recession fails to produce a plausible answer. Why? One possibility is that early on, the Moon's orbit was altered by a near-passing celestial body, perhaps a "rough planet." Another is that its orbit was influenced by the gravitational pull of "dark matter" (a material that comprises roughly 25% of the mass of the universe and that, because it does not emit light or energy, cannot be directly observed). Theories abound, but at present there is no direct evidence, no way to know for certain.

The history of Earth's day-length over geological time is one of several major problems for which there currently is no firm answer. How can its solution be found? Unfortunately, studies like those of John Wells of the daily growth increments of invertebrate animals (corals, brachiopods, and the like) cannot be used in the Precambrian Era (nearly 90% of Earth's history) because such organisms had not yet evolved, first appearing about 550 million years ago during the era-ending "Cambrian Explosion of Life."

Nevertheless, it seems possible that the telling evidence may be encoded – albeit in an abbreviated form – in the genomes, the genetic material of living systems. All photosynthetic forms of life, microbes included, are cued into their daily environment by circadian rhythms, genetic clocks that turn on and turn off photosynthesis on a regular day–night cycle. As day-length has changed, the internal genetic clocks would have changed as well. And as more and more becomes known about the workings of these clocks – and if their present-day cycle can be excised and their original cycle can be reconstructed and then linked to the fossil record to show their time of emergence in the geological past – changes in day-length over time might be revealed. If successful, such studies would document the gradually evolving workings of Earth's global ecology and the history of the Earth–Moon system as well.

There are numerous other problems that beg for just such an integrated genomic-fossil record-based approach, among them the interrelated history of ultraviolet radiation and life. Based on understanding of the "life histories" of main sequence stars such as the Sun, the solar-generated UV-flux reaching the Earth should have decreased markedly over early geological time, a change to which life would have adjusted. Here the prediction would be that intracellular mechanisms for the repair of UV-caused damage would have decreased in newly evolved primitive lineages in parallel with the decreasing UV-flux and that they later would have decreased even more dramatically with the mid-Precambrian advent of an oxygenated atmosphere and its accompanying UV-protective ozone shield. Evidence of such changes should be discernible in the genomes of modern living systems with the timing and rate of change of the mechanisms to offset UV-damage being revealed by the fossil record–documented time of the evolutionary appearance of the relevant branches of the Tree of Life.

9.12.2 Evidence of the "Primordial Soup"

Another vexing unsolved problem that falls in the purview of Precambrian Paleobiology is the complete absence of direct evidence to establish the existence of an Oparin-Haldane-Miller-Urey-et al. non-biologically produced organic-rich "primordial soup" in which life is thought to have originated and then evolved. There are numerous competing ideas as to the local setting in which life got started, ranging from Darwin's "warm little pond" to open ocean-shallow bays-lagoons, to deep-sea volcanic fumaroles. But all require the non-biological production of the organic building blocks of life – all demand the prebiotic formation of the organic compounds required for life to begin – for which there are good laboratory-simulated models but no known direct evidence from the geological record.

As an explanation for this absence of firm understanding, one might turn to the known continuous rock record, which extends only to some 3,500 million years ago when abundant fossil evidence shows that life had already emerged. Life's beginnings must date from earlier, perhaps 4,000 million years ago, but rocks of that age have been eroded away, leaving no firm evidence to investigate. While this is the current situation, by the end of this century, new telling finds may become available as global warming melts the two-mile-thick Greenland icecap to expose rocks approximately 3,800 million years old, which we know exist but cannot yet properly sample.

There may be other lines of evidence that could now be exploited. Prebiotic *non*-biological syntheses of life-generating organic building blocks require only three main components: (1) biogenic element- (carbon, hydrogen, oxygen, nitrogen, sulfur, and phosphorous [CHONSP]) containing starting materials (available on the early Earth as they are throughout the Cosmos); (2) an energy source to initiate the necessary chemical reactions ("starlight," ultraviolet radiation from the Sun, which early in Earth history was appreciably stronger than today); and (3) an absence of atmospheric oxygen, which would have destroyed such products before they could accumulate (but which would not have been plentiful before the origin of oxygen-producing photosynthesis and the later build-up of oxygen in the mid-Precambrian Great Oxidation Event).

Thus, based on current understanding, there ought to be evidence of such non-biologically produced organic matter – regardless of whether it was formed by atmospheric Miller-Urey syntheses or by Fischer-Tropsch-type syntheses in submarine fumaroles at the ocean bottom – sequestered in the rock record up to the mid-Precambrian rise in environmental oxygen. Unfortunately, however, such abiotic products would not be easy to detect since they would have made up only a minuscule component of the preserved organics, swamped by the vastly more abundant carbonaceous matter produced by Earth's thriving Precambrian microbial biota.

How might such an almost vanishingly small signal of *non*-life organics be detected? This is a tough problem, no doubt widely recognized (though it is almost never seriously addressed). Clearly, the minute signal would be expected to be sequestered in the tiny bits and pieces of the coal-like carbonaceous kerogen sequestered in ancient rocks. To date, however, much of the work in unraveling the detailed chemistry of such organic matter has used the technique of pyrolysis-mass spectrometry, excellent for some problems but not for this one, because the technique itself makes new (assuredly non-biological) products that are not present in the original material.

Nevertheless, at least in principle, all may not be lost. Coal-like carbonaceous kerogen is composed of layer upon layer of polycyclic aromatic hydrocarbon (PAH)–dominated planes (called "graphene layers"). The PAHs and the stacked layers are internally and vertically linked together by short bridges of carbon- and hydrogen-rich organic compounds. Such bridges, unlike the PAHs, potentially encode biological (as well as non-biological) information. The problem is to excise these bridges and decipher their origin.

Biological organic matter is formed by biochemical reactions mediated by enzymes that first produce small molecules, "monomers," that are then commonly joined together head to tail to build larger and larger compounds ("polymers"). For example, the biosynthesis of the 60-carbon polymer squalene (which, with subsequent modification, leads to the formation of the human gender-determining steroid hormones estrogen and testosterone) begins with an enzyme-produced small five-carbon molecule. This monomer is then linked to another to make a 10-carbon pair, and this pair is then joined to another 5-carbon unit to make a 15-carbon chain. Two 15-carbon chains are then linked together to make a 30-carbon unit, and two of the 30s are finally linked to make the 60-carbon squalene hormone-precursor.

Because such biosynthetic processes are midwifed by enzymes that carry out precise chemical reactions, the isotopic composition of the carbon atoms emplaced in the original subunits can vary systematically carbon to carbon. For example, the thousands of six-carbon glucose sugar monomers of the cellulose that makes up the cell walls of plants (and the paper on which this book is printed) are reported to have the heavier carbon isotope ^{13}C at the beginning (C-1) and ending (C-6) positions of each and every subunit, with isotopically lighter ^{12}C interspersed between them. In contrast to such enzyme-directed formation, non-biological syntheses are essentially random, so their products would not have such a regular pattern of intramolecular isotopic distribution.

So far, so good – a biological vs. non-biological signal can be identified. But how can the evidence-containing snippets of the coaly kerogen be fished out for analysis? Here the answer may lie in genetic engineering. Though relatively few hydrocarbon-snipping microbes occur in nature, some are known and others have been bioengineered to clean up oil spills. So, it ought to be possible to design a microbe that will cut apart the short bridges that link together the graphene layers of fossil kerogens and to then concentrate and analyze the freed bits and pieces to determine their patterns of intramolecular isotopic distribution. Most would exhibit a carbon-isotopic regularity, evidencing their enzyme-driven biological synthesis. But a smattering of the others might not. After whatever other possible causes of such lack of regularity were ruled out, these rare tiny snippets, perhaps one in 10,000, could plausibly be regarded as being products of non-biological organic syntheses, evidence of a prebiotic "primordial soup" like that in which life originated.

Studies of this sort might even be required for studies of Mars rocks. If the samples to be cached by NASA's Mars2020 mission and then brought to Earth are shown to contain kerogen-like organic matter, its source will be fiercely debated. If it truly formed on Mars – and especially if it is widespread and abundant – it might well evidence past Martian life. On the other hand, as is bound to be suggested, it might actually have been implanted on Mars by an impacting carbon-rich meteorite, in which case the organics would be solely *non*-biologic, synthesized in space by reactions of CHONSP and starlight. Studies of the intramolecular isotopic composition of such kerogen-like organic matter could reveal its source.

9.12.3 Oxygen, Nitrogen, Energy, and Life

Life needs energy to live, to grow, to reproduce. And among the various rules of biological evolution one stands out: In the competition among similar cellular *energy-producing* processes, those that generate a larger amount of energy win, and those that are less energy-yielding lose. This explains why oxygen-using (i.e., aerobic) plants and animals have become so prevalent – by the process of aerobiosis, "breathing," they generate about 16 times more energy from metabolizing a single molecule of glucose sugar than did Earth's earliest primordial soup-eating anaerobic microbes (which relied solely on the more primitive process of fermentation). The same should be true in reverse: In a competition between *energy-consuming* processes, the more efficient *less energy-requiring* forms should win and those that use *more energy* should lose.

There are, however, exceptions to such energy-based evolutionary rules. The oxygen-producing photosynthesis of cyanobacteria requires more energy, not less, than its *non*-oxygen-yielding evolutionary precursor – yet the fossil record shows that cyanobacteria, inventors of the process of O_2-producing photosynthesis (used also by algae and higher plants) were nevertheless highly successful, the dominant primary producers of Earth's Precambrian biota. Similarly, the microbial process of nitrogen fixation, the capture and use of atmospheric N_2 for its incorporation into the amino acids of enzymes, is vastly more energetically costly than the earlier-evolved use of ammonia (NH_3) or nitrate (NO_3), yet nitrogen fixation originated early in life's history and has survived to the present (as, for example, in the root nodules of rice and other crops).

Let's first deal with the oxygen story, which is particularly interesting. It costs quite a lot of energy (technically, 111 Kilocalories/mol = 464 Kilojoules/mol) to break oxygen away from a molecule of water (H_2O) in oxygen-producing photosynthesis – in chemical parlance, $CO_2 + H_2O + \text{light energy} \rightarrow$ "CH_2O"+ O_2. This is appreciably more than the energy required (81 Kcal/mol = 339 Kj/mol) to free sulfur from the hydrogen sulfide (H_2S) used in the more primitive non-oxygen-producing form of photosynthesis ($CO_2 + H_2S + \text{light energy} \rightarrow$ "CH_2O"+ S). Thus, the highly successful oxygen-producing photosynthesis of cyanobacteria costs more energy, not less, than its more primitive *non*-oxygen-yielding evolutionary precursor.

Despite this difference in energy requirements, O_2-producing photosynthesis won, while the more primitive, albeit less energy-using and earlier evolved form, lost. Why? In part it may have to do with the relatively limited availability of the hydrogen sulfide gas needed for the more primitive form of photosynthesis, large supplies being centered only near H_2S-belching volcanic fumaroles. In contrast, the water needed for O_2-generating photosynthesis was more or less ubiquitous across the planet.

But there is another, perhaps even more compelling, plausible explanation for why O_2-producing photosynthesis won the battle. The earliest cyanobacteria, the first O_2 producers, were mutants of their non-oxygen-producing evolutionary precursors, photosynthetic bacteria. In the competition for the sunlight needed to power photosynthesis of both of the two strains, the cyanobacteria and the photosynthetic bacteria were in cheek-by-jowl combat. Yet the oxygen given off by the cyanobacteria was a deadly poison to their earlier-evolved competitors. Indeed, gaseous oxygen was a completely new addition to the environment; all earlier evolved forms of life (primitive sulfate reducers, methane producers, methane consumers, and everything else) had never before been forced to cope with it. *Gas warfare!* The earlier-evolved photosynthetic bacteria had but one choice – retreat or die – and some types did in fact retreat, finding an oxygen-free safe haven beneath the surface layer of cyanobacteria, where they could absorb parts of the Sun's energy (for example, in the infrared region of the solar spectrum) that seeped down to their oxygen-free zone unused by the overlying oxygen producers.

This story makes good sense and is probably correct, but its timing is difficult to nail down in the fossil record. Based on the suggestive evidence Bill Schopf has seen, his current guess is that O_2-producing cyanobacteria may have taken over Earth's ecosystem at about 3,000 million years ago. But this is only an "educated guess," and in science, guesses and speculations do not count. However plausible this timing may seem, it cries out for supporting facts.

Nevertheless, if this overall scenario is more or less correct, there must have been a very long segment of geological time, nearly a billion years, between the origin of O_2 production and the eventual increase of atmospheric oxygen during the Great Oxidation Event (GOE) about 2,200 million years ago. During this pre-GOE period, the rock record shows that oxygen levels were kept low by it being sponged out of the atmosphere by chemically combining with gases given off by volcanoes and by being removed from upper reaches of the ocean by its reaction with dissolved iron (Fe^{+2}) to form insoluble particles of rusty iron oxide (the mineral hematite, Fe_2O_3), which rained down on the ocean floor to produce the deep-red iron-rich rocks we mine today to make steel. Goodness gracious – the world rusted for a billion years!

During this billion-year period, the rock record also shows that both short- and longer-term "whiffs" of oxygen occasionally built up, chiefly during periods when volcanic activity (and the O_2-sponging gases it gives off) was relatively low. And from place to place, where cyanobacteria were actively producing oxygen, there must have been "oxygen oases" where during daylight hours microbial O_2 breathers would have flourished. Thus, during this exceedingly long period, Earth's surface-dwelling biota would presumably have been populated by swarms of facultative aerobes, "on-and-off" oxygen consumers able to use oxygen for energy- production when it was available but, in its absence, switching back to the less effective energy-yielding process of fermentation (anaerobic glycolysis).

Many such facultative "switch-hitting" microbes are known today, thriving at the oxygen-rich/oxygen-poor interface (the oxic-anoxic "chemocline") in the shallow regions of some lakes and more widely just beneath the surface of bottom-deposited mud. But such microbes have yet to be identified in the fossil record, and the possible use of the gene-based techniques of ancestral molecular sequence reconstruction ("ASR") to trace the history of such facultative O_2 use back through time has yet to be investigated.

The origin of microbial nitrogen fixation is another piece of the early-life puzzle for which we have a plausible story but only scanty evidence. The two nitrogen atoms in nitrogen gas (N_2), some 79% of Earth's present-day atmosphere, are tightly tied together by three strong bonds that in total require a whopping 226 Kcal/mol (945 Kj/mol) to break apart (a linkage said to be the strongest bond in nature). For microbes to break this bond, to free the nitrogen atoms for their incorporation into the amino acids of enzymes, requires the expenditure of 12 energy-rich molecules of ATP (adenosine triphosphate), an exceedingly high cost that evolution would not have invented unless this process had been required for survival. But microbial N_2 fixation is widespread among the deepest, most ancient surviving early-evolved branches of the Tree of Life. This, in turn, suggests that there must have been a "nitrogen crisis," a scarcity or near-absence early in Earth history of the nitrogen needed to build the amino acids required by living systems.

What sources of nitrogen were available on the early Earth? Certainly there must have been ammonia, NH_3, but there was also a high ultraviolet flux from the early Sun. Calculations made years ago by Philip H. Abelson (famed for his contributions to the Manhattan Project and the former editor of *Science*) show that if all of the 79% of the nitrogen in today's atmosphere were converted to ammonia, the first choice of modern N_2 fixers, it would have been UV-destroyed in a scant 1,000 years leaving as its only major source the recycled nitrogen produced from the decay of dead microbes. The second choice of living N_2 fixers is nitrate, NO_3. In today's world, most of the biologically available nitrate is produced by the reaction of atmospheric N_2 and O_2 caused by lightning – but because oxygen was scarce in Earth's atmosphere, this source would also have been minimal. Thus, use of atmospheric N_2, the last resort of modern N_2 fixers, would have been crucial to the production of the amino acids needed to build enzymes – and since enzymes power the biochemical reactions of living systems this would have been critical to the survival of life itself.

As Theodosius Dobzhansky (1900–1975) taught us in 1973, "*Nothing in biology makes sense except in the light of evolution.*" Dobzhansky's insight is assuredly correct and, given the current state of knowledge, our stories about oxygen, nitrogen, life, and evolution all seem to make sense. But such understanding also features components that border on being "just-so stories" and "by-guess-or-by-golly" notions that may reveal more about present-day ignorance than well-established knowledge. The times of origin, immediate impact, and histories both of O_2-producing photosynthesis and N_2 fixation have yet to be properly documented, problems for which firm data are needed from the Precambrian fossil record backed by the findings of Biomarker Geochemistry and Genomic Biology.

9.12.4 Microbial Consortia

As surprising as it may seem, environmental microbiologists estimate that fewer than 2% (with some estimates dipping as low as 0.1%) of the myriad types of microbes that occur in nature can be grown in laboratory cultures. And what that means is that we do not know much about them. It should be noted that "not able to be cultured" merely indicates that the vast majority of such microorganisms cannot *currently* be grown and studied under laboratory conditions – not that they "can never be cultured" – and it is certain that many such strains will be cultured in the future. Nevertheless, this present-day inability to carry out detailed studies of such microbes is a striking if disheartening acknowledgement of the current state of ignorance. What are the roots of this lack of knowledge and how might it impact understanding the Precambrian history of life?

A part of this problem stems from a lack of knowledge and the resulting difficulty of replicating the tightly constrained, precise environmental conditions certain types of microbes require for growth. But an appreciably broader more pervasive difficulty is that such microorganisms commonly live symbiotically as interacting members of microbial consortia (as, for example, in sewage treatment ponds and in the intestines, the "guts" of everything from worms to cows to humans). The term *symbiosis* refers to the close association between two or more organisms of different species, members commonly of decidedly differing evolutionary lineages. And if the association benefits each of the interacting members it is referred to as a *mutualistic symbiosis. Microbial consortia* are naturally occurring assemblages of such microbes, each type interacting with, and in some cases being wholly dependent on their interactions with, one or more of the closely coexisting types of microbes.

The fossil record of microbial consortia is as yet very poorly known. Indeed, the very first such consortium was reported by Bill Schopf and his colleagues in 2015 – only three years ago – mid-Precambrian (1,800- and 2,300-million-year-old) "sulfuretums" in which one microbial component generated hydrogen sulfide gas and a second used this H_2S to produce energy. Since then, in 2017 and 2018, they ferreted out two more such assemblages. What is known now from these three examples is only a minuscule tiniest tip of a monumental and absolutely important underlying iceberg!

As a simple two-member example of a mutualistic symbiotic microbial consortium, consider the interactive relationship between sulfate-reducing bacteria and non-oxygen-producing (anoxygenic) photosynthetic bacteria. Both of these very different types of early-evolved microbes are anaerobes, organisms that thrive in the absence of oxygen, so they can live together in the same oxygen-lacking environment. The major difference between the two is that the sulfate reducers can live in the dark whereas the photosynthetic bacteria need light to power their photosynthesis. This mutualistic association wouldn't happen today at the light-flooded surface of the Earth because that environment is rich in oxygen. But on the early Earth, before the development of an oxygen-containing atmosphere, they could have lived together in mutualistic harmony exposed at Earth's surface in shallow lagoons and even in intertidal mudflats.

The sulfate reducers take in oceanic sulfate, combine it with hydrogen from decaying carcasses and give off hydrogen sulfide, the foul-smelling gas of rotting eggs (in shorthand, $SO_4 + H_2 \rightarrow H_2S$). All goes well if the H_2S diffuses away from the sulfate reducers into the environment or if it is sponged up by its reaction with unoxidized iron to produce iron sulfide, the mineral pyrite, known also as "fool's gold" ($H_2S + Fe^{-2} \rightarrow FeS_2$). But if the H_2S builds up locally and engulfs the sulfate reducers – when it is captured, for example, in the sticky mucilage sacs that often encase microbial assemblages – it is lethal. The sulfate reducers die. (Interestingly, exposure to too much H_2S is also deadly to humans – we die from asphyxiation. Thanks to our evolutionary heritage, however, our noses can detect even minute concentrations of noxious H_2S gas from which we and other mammals recoil. HA! Being able to detect this stinky gas is rooted in our genes!)

How can sulfate reducers avoid killing themselves in a mass suicide if they are encased in mucilage? One solution would be to turn off their intracellular H_2S-producing machinery. But this they cannot do because this is the very process that powers their cells, and they soon would then die anyway (like a human who refuses to eat). That won't work and it is not programmed in their genes, the First Rule of Life being to "*stay alive!*" To defeat this problem, their compatriots in this simple two-part microbial consortium, the photosynthetic bacteria, come to the rescue. These bacteria, the earliest evolved photosynthesizing microbes, take in hydrogen sulfide and carbon dioxide and, powered by light energy, use them to make sugar: $H_2S + CO_2$ + light-energy → "CH_2O"+ S. The photosynthetic bacteria consume the lethal H_2S. Voila! The problem is solved by symbiosis!

It is worth noting that neither the sulfate reducers nor the photosynthetic bacteria in this mutualistic symbiotic association can properly be termed "altruistic" – that is, each is simply "doing its own thing," the effluent of one being used to power the photosynthesis of the other. Neither is motivated to purposefully "try to be nice" to the other, their gene-driven imperative being to take care of themselves, not some other tribe of microbes. The photosynthetic bacteria are attracted to the sulfate reducers as a handy source of the H_2S needed to power photosynthesis, and the sulfate reducers benefit because the photosynthesizers sponge up their H_2S effluent, which, in large amounts, can to them be lethal. Thus, exposed at the surface of the early Earth – where in the absence of atmospheric oxygen the two could live in harmony – these two very different groups of organisms composed a mutualistic symbiotic microbial consortium, a symbiosis in which each benefited, survived, and thrived.

This is a nice story and it is evidently correct, as Bill Schopf and his colleagues showed in 2017 for just such a 3,400-million-year-old mudflat-inhabiting microbial consortium. But, surprisingly, this was only the second example of a microbial consortium to be identified in the geological record.

In 2018, other colleagues and Bill documented yet a third microbial consortium, the symbiotic relationship between the methane-producing and methane-consuming microbes of the nearly 3,500-million-year-old Apex chert discussed in Chapter 8. Here, the symbiosis was a bit different. Instead of being a *mutualistic* symbiotic microbial consortium, this one is more properly dubbed *commensalism*. What that means is that unlike a mutualistic "I scratch your back, you scratch mine" both-sides-benefit situation, this commensalistic symbiosis was "one-sided." In particular, *commensalism* refers to a symbiotic relationship between members of two disparate species in which one (the "commensal" species) obtains benefits (e.g., nutrients or protection from predators) from the other without affecting the other (the so-called "host species"). In this example, the methane consumers (the commensals) obtained the benefit of the effluent methane gas produced by the co-existing methane producers (the hosts).

But when it comes to microbial consortia known from the fossil record, *that's it!* Only three examples from the entire geological record – all reported in just the past three years. Yet the living world is replete with such consortia, with the evidence now showing that such associations have existed for literally billions of years. We know hardly anything about such interactive microbial communities, whether living or fossil. With all this yet to be learned, the future holds great promise!

How will this aspect of the science advance from here? The answer seems obvious. It is not sufficient for Precambrian paleobiologists to study only the morphology, the shapes and forms and taxonomic assignments, of ancient microbes. There are a few "stand-alone" types of free-living ancient microbes, such as cyanobacteria, for which there is a reasonably well-known fossil record. But such "free-livers," despite their ecologic dominance in particular environments, represent only a minuscule fraction of the totality of microbes that flourished on the Precambrian planet.

What is needed for the science to advance is an interdisciplinary approach, a "professional symbiosis" between paleobiologists and environmental microbiologists. In the coming decades, microbiologists will be able to harvest an increasingly larger percentage of microbes from nature that can be cultured and studied in the laboratory, and their growth requirements and consortial associations will become increasingly better understood. This knowledge should then be used by paleobiologists to guide their search for ancient microbes and microbial consortia in the fossil record. As the discoveries of the past few years well show, this approach has and will bear fruit. In short, the findings of modern microbiology can provide invaluable grist for the mill of those ferreting out life's earliest history.

9.12.5 Rates of Evolution

It is certain that some branches of the Tree of Life evolved relatively rapidly whereas others developed far more slowly, a difference referred to in genome-based phylogenies as "fast evolving" and "slow evolving" lineages. Evidence of these vastly varying rates of change and an explanation of why they differ so greatly ought to be sequestered in the information encoded in the genetic material of modern organisms – but as yet we do not know how to accurately decipher such clues.

At present, the fossil record provides the only meaningful evidence, for example by documenting the rapid evolution that occurs after the development of a new body plan (e.g., during the Cambrian Explosion of animal life and the later rise of plants on land) and the slower evolution, the long-term stasis that so frequently follows such episodes, as Earth's ecosystem came to be stabilized.

But the fossil record is not without deficiencies, simply because fossils can only document the oldest *known* appearance of a given branch the evolutionary tree rather than the *actual* time that lineage came into being. Think about it. The odds of any type of organism being fossilized depend greatly on its abundance. Rare forms of life living in small populations are almost never preserved as fossils, whereas the great abundance required for preservation indicates great success. Such success, however, typically requires a considerable period of time. As a result, the "oldest known occurrence" of any branch of the Tree of Life can be appreciably younger than its actual time of origin. Over the present century, as we learn more and more about the genetic bases underlying the fossil-evidenced evolutionary progression, a reliable molecular clock for the deep-time history of truly ancient Precambrian life can be expected to be devised.

9.12.6 A Paleobarometer and Paleothermometer

There are numerous other major unanswered questions about life's Precambrian history, for which at least two a good start has been made.

The first of these centers on the history of the biologically crucial increase of atmospheric oxygen over the geological past, addressed by the construction of a mineral-based quantitative O_2 paleobarometer that has thus far been applied only to the time spanning the Precambrian to Cambrian boundary and the Cambrian Explosion of Life. And it works. Based on the oxygen-sensitive substitution of the rare earth element samarium into the crystal lattice of the mineral apatite, a reliable paleobarometer has been constructed that produces excellent results for low concentrations of locally available oxygen. What is needed, however, is a tool such as this that covers the entire 0%–21% (and on occasion even greater) spectrum of the oxygen content of Earth's atmosphere that can be applied over all of geological time.

The second such partly resolved question is the history of Earth's surface temperature over geological time, the most recent telling results coming from the use of the thermal stability of reconstructed ancient enzymes. Spurred by earlier isotope-based analyses of marine cherts, which postulated a so-called "Hot" (~75°C = 167°F) early Earth – an interpretation not widely accepted because of its misfit with long-held assumptions – the entirely independent enzyme-based findings seem especially promising. Studies of ancestral forms of the biologically ubiquitous enzyme NDK (nucleoside diphosphate kinase) in living photic-zone cyanobacteria, algae, and land plants, combined with the known Precambrian fossil record, reinforce the concept that Earth's surface temperature has gradually decreased over time. Such work, however, has only just begun. Data from other enzyme systems, supported by more and better data from the fossil record, will be needed to confirm these promising mutually reinforcing geochemical–genomic recent findings.

There are, of course, a great many other yet-unsolved problems about the interrelated early history of life and the environment, some now known and others that will later arise. Precambrian Paleobiology is a new science, still only 50 years young. As time marches on and the science continues to mature, more and better data will be amassed. That is the way of science, the very best "detective game" we humans have ever devised in its long-term, ongoing quest to describe and understand the real world and its vast history.

10 Final Comments

10.1 DÉNOUEMENT

Having defined his career path early in life, over the following span, J. William (Bill) Schopf gained experience in Geology, Paleontology, Paleobotany, Ecology, Microbiology, Biochemistry, Genomic and Evolutionary Biology, Organic and Isotopic Geochemistry, Atmospheric Evolution, and Comparative Planetology. And like all serious academics, over his career he has tried to use this understanding to contribute usefully to his fields of special interest (Precambrian Paleobiology, Astrobiology, and Evolutionary Biology), to his university, and, more broadly, to human knowledge.

Over this period, Bill has grown intellectually as he matured from an enthusiastic and well-meaning (if assuredly unseasoned) fledgling academic to an elderly, more knowledgeable, and hopefully wiser member of the professorial guild. At the end of each decade of his life – when he turned 40, then 50, then 60, then 70 as the years sped by at an ever-quickening clip – he has found it useful to (privately) review his performance and ask himself "Billy, you aging guy, are you any good?" He has yet to be satisfied. Oh yes, he likes to think that he has made progress – he imagines that he has improved – but he is first to acknowledge that he still has a great deal to learn.

Nevertheless, Bill has grown past what he now supposes to have been a fortunately relatively brief early phase of "recently-minted-Ph.D.-me-firstism" to become a confirmed group-first interdisciplinary internationalist, increasingly focusing over the years on the needs of others – students, his department, the University of California Los Angeles (UCLA), fellow-workers in his science – and using his acquired expertise to help many workers in many countries. And he has traveled the world – England and Europe; China; India; Japan; Russia; North, Central, and South America; parts of the Middle East; South Africa and Kenya; and the South-Pacific nations of Australia and New Zealand. To Bill, these experiences have been extraordinarily helpful, teaching him to appreciate the history and cultures of other peoples and to realize that "the American way," as much as he values, respects, and enjoys its remarkable benefits, is not the *only* way to wend one's way through life.

As you will have gathered from this narrative, Bill Schopf is a happy, "upbeat" academic, grateful for having such remarkably gifted colleagues and friends who have made his life so richly rewarding. To some, one might suppose, his life and career may seem to have been "one of a kind." After all, there are not many among us who have had the opportunity to witness, participate in, and then take on a significant role in helping to forge a game-changing Kuhnian "paradigm shift" in human knowledge. But Bill knew even as a grade-school youngster that he was a bit different, that for some inexplicable reason he had a capacity to both think through and to seek the roots of the problems he encountered and the "courage" (or unwitting immaturity) to carry his notions to fruition – traits illustrated, for example, by his youthful "Chicken Chase" and the "Break-in" to his grade-school gymnasium.

Like all of us, of course, young Bill always wanted to fit in with his friends, to be part of the group – think, for example, of his high-school days, when he scrubbed and waxed floors to buy an up-to-date pink shirt to be like his pals. And such "fitting in" was for him never much of a problem, whether in sports, music, schoolwork, paid employment, even with girlfriends. But already as a youngster, he had hopes and dreams that differed from his classmates, announcing to his 4th-grade class that he aspired to be a "professor." He didn't "compute." And it is this individuality, coupled with Bill's curiosity, problem-solving penchant, and dogged persistence, that has carried him through his life in science.

Over the years, Bill has managed to embrace these traits, convincing himself that they are a positive, not a negative. And over his life, he has continued, undeterred, to trundle on and "do his thing" – as, for example, when he broke a 50-year precedent by becoming the first Geology Honors Student in the history of Oberlin College, despite the lack of faculty guidance for the project he carried out. Moreover, he has been richly rewarded, being the recipient of all campus-wide awards available to UCLA faculty – for teaching, research, and academic excellence. Over his career he has also received three national book prizes; seven medals from national and international scientific societies; an honorary doctorate degree from the University of Pretoria, South Africa; and has been elected to membership of all of the United States' most prestigious scholarly honorific societies. To him, understandably, his life has been wonderful great fun!

Given all that, the concluding passage of this narrative is Bill's own assessment of his career and his now nearly 60-year-long quest to unearth the record of early life on Earth, the 85% of life's long history that, until his college days, had eluded science for the entire preceding century, a first-person account that reveals what he thinks about his life in science and how this field has unfolded.

As I now think back on my life, there are five main pillars that have held me in good stead, observations and lessons learned that may be of use to others: (1) the sheer joy I find in having a fresh idea and carrying it to fruition; (2) honesty and integrity, without which science could not function; (3) decency and openness to others, a prerequisite for positive human interactions; (4) an interdisciplinary understanding of the natural world, the foundation of the science I have tried to do; and (5) internationalism, an appreciation of the history, traditions and values of other cultures – one of the reasons that I so greatly value UCLA, where we are so wonderfully fortunate to have in our midst such a remarkable diversity of faculty and students, a stunning representation of the varied "flavors" of our world's population.

As you *look back on my life and my science, you are all too likely to give me too much credit.* Please do not make that error! *If in your eyes I have accomplished anything of merit others deserve the lion's share of the credit, not I. Whatever I may have done of note is fundamentally a product of my heritage (genetics); my upbringing (my parents, my brother, my teachers, my schooling); my interdisciplinary liberal arts education at Oberlin College; my mentor in science and graduate professor, Elso Barghoorn; my UCLA fellow faculty members and generous supportive students; my professional colleagues worldwide, particularly the members of the Precambrian Paleobiology Research Group; and – above all else – my dear wife Jane, a wonderfully intelligent, wondering person and the fundament of my existence. So, if credit is due,* bestow it on them!

The oft-repeated aphorism coined by Louis Pasteur that "chance favors only the prepared mind" has for me occasionally held true: first, when I set out on my life-long quest to discover the so-called "missing" early history of life, knowing that Darwin was right and that the then century-old dogma made no sense; second, decades later, when I discovered promising evidence in two-billion-year-old rocks that support the required "null hypothesis" of Darwinian Evolution, a notion I had come up with in a Philosophy course as a 1st year college student 50 years earlier; third, after some 40 years of futile search, when I finally found ways to solve three problems basic to my science that had niggled away at me since my graduate-student days, methods to analyze directly, in situ, *the 3-D morphology and both the molecular and isotopic compositions of tiny microscopic fossils completely embedded within rocks; and fourth, more recently, when I have sought to combine the evidence from microbiology and genomic biochemistry (from my biology background) with the findings of geochemistry and paleontology (my geology training) to document the metabolic characteristics of Earth's earliest biota and the temperature of the planet's near-surface environment over billions of years of geological time.*

Interestingly, and pretty obviously, over my career I have experienced two phases of particularly useful creativity: (1) when I was a young questioning student, too ignorant and inexperienced to know, much less necessarily accept, the then-current dogma; and (2) later, when I was much older and had finally acquired the multi-disciplinary knowledge and experience needed to more fully analyze the pressing questions of my field and to devise new ways to attack the problems that they posed.

Nevertheless, it remains clear to me that over my entire life, I have simply been an exceedingly lucky guy – *in the right place, at the right time, interacting with the right people, learning the right stuff, asking some of the right questions, and having the perseverance to plow ahead and do the right work – this last, a function of what my mother called "stick-to-itiveness." Honest to Goodness, time and time again, wonderfully nice things have happened to me, over which I have had not the slightest control – and many of which, quite frankly, I hadn't earned and really didn't deserve.*

Why has all this good stuff happened to me, whether in music, or sports, or school, or teaching, or research, or personal life, or in my professional career as I have tried to move my science to greater heights? Without doubt, luck has played a significant role as has, on occasion, Pasteur's notion of a "prepared mind." But underlying this, I think, is a far more basic answer: the values *embedded in me from my childhood, derived largely from my father and, through him, from my Mennonite grandfather and our protestant Calvinist Huguenot ancestor, Christian Schopf – like the ancestors of a great many Americans, a talented but unrewarded "misfit" in his home country – who found himself forced to immigrate to this country in the early 1700s to escape religious persecution and who had the courage to embark on a path toward a better life. It is his values, his fundamental principles – for me, nearly three centuries later – that, devoid of their religion-based theistic aspects, have guided me throughout my life. For this, I am sincerely appreciative.*

I have, of course, failed – far too often. Like all of us, I am by no means "perfect." Indeed, I have made numerous errors, most prominently in my interactions with others and my inability to see myself as others see me – I am a lousy mind-reader! I know that I have tried mightily to do what I thought was right, but due to my own ignorance and/or carelessness, I have often blundered.

Nevertheless, because of my values, I have done my best to learn from my mistakes, both trivial and more serious. In fact, virtually every night, as I hunker down to sleep, I privately quiz myself: "How did the day go? Could you have done better?" I am virtually never satisfied with my performance. "Good enough" is simply not good enough. And even when I am really pleased with my day – those happy times when I imagine that, like in my grade-school days, I had hit a home run with the bases loaded – my prime question is "Can you do it again tomorrow, and the next day and the next?" To some, this would be a burden. But it is not so for me. Though the bar I set is high and I almost never measure up, I keep trundling along, I keep trying to do as best I can. It is a challenge that I am determined to meet.

My guiding principle is to believe in myself and – far more importantly – *to believe in the goodness of others and to help them to the best of my ability. Throughout my life, I have tried to follow the mantra, "If you see a problem, solve it; if you see a wrong, right it." It is for reasons such this that my students and I have introduced new field-changing analytical techniques – problems we identified and then managed to solve – and why it is that I have devoted effort to correct the deleterious effects of a myopic one-dimensional rather than interdisciplinary approach to the game-changing science to which I am devoted.*

Simply put, I urge myself to charge ahead and do it! *And I try never to blame someone else for my failures or misdeeds. When I fail, when I have fouled up – just as I did when from my stupidity of looking down to the bottom of the gorge when my legs went out of whack and I was marooned on a precipitous cliff in Western Australia, or when I again stupidly brought the wrong set of slides to my maiden talk at the American Philosophical Society – I know full well that it's my own darn fault!*

In sum, I really do think that values *are the fundament of life, and it is difficult for me to imagine that anyone could ever have been as fortunate, as lucky as I. Moreover,* I cannot imagine – and do not believe – that there is anything all that "special" about me. *If my overriding principles work for me, as my past success shows they have, it would seem to me that they might well work for others, too.*

If that foregoing (honest, if assuredly self-revealing) harangue is a bit hard to stomach, please understand that I was born a scant three months before the attack on Pearl Harbor and FDR's even-now memorable pronouncement: "December 7, 1941 – a date which will live in infamy." So I am an "old guy," my formative years spent in the post-World War II period of the Marshall Plan and national and international recovery. Pause for a moment and think of the movies of those days (all in black and white) that always had a happy ending … and if you think of me, think of Jimmy Stewart in It's a Wonderful Life. *But if that doesn't explain me to you as a happy, "upbeat" person, grateful for having experienced such a wonderfully rewarding life, think of the lyrics of Frank Sinatra's 1969 ballad "My Way," which might, possibly, help you understand that even though I knew as a grade-school youngster that I did not "compute" and never would, to me it has never really mattered.*

OK, as applied to me, the lyrics of "My Way" are way over the top. Sinatra was a Hoboken, NJ idol of the zealous "Bobby Soxer" crowd, whereas I am merely a Midwest-derived UCLA academic. Still, I would like to think that most of the lyrics ring true, not just for me but for the paradigm-advancing Precambrian Paleobiology Research Group (PPRG) as well and its monumental accomplishments. So, with apologies to Sinatra and lyricist Paul Anka, here is my re-titled, recast version of their ballad, intended as a much-merited homage to the founding heroes of this science and the game-changing 50 members of the PPRG.

Our Way

And now for me the end is near
as I face the final curtain.
My friends, I'll say it clear,
I'll state my case, of which I'm certain.
I've lived a life that's full
and traveled each and every highway.
And more, much more than this, *together we did it our way.*

Regrets, I've had a few,
but then again too few to mention.
We did what we had to do,
and saw it through without exception.
We planned each charted course,
each careful step along the byway.
And more, much more than this, we *did it our way.*

Yes, there were times, I'm sure you knew,
when I bit off more than I could chew.
But through it all,
when there was doubt,
we together ate it up and spit it out.
We faced it all and we stood tall,
and *did it our way.*

I've loved, I've laughed, and sometimes cried
with ups and downs, and I'm not teasing.
And now as life fades on, it is so very pleasing
to think together we did all that.
And may I say, not in a shy way –
Oh, no, oh, no, not me –
we did it our way.

For what is a group
and what's it got?
If not its values, then it has naught.
To say the things we truly feel

and not the words of those who kneel.
The record shows we took the blows
and *did it our way.*

Yes, it was Our Way!

My task of preparing this essay is now completed – I have done what I can to meet the urging of Tom Kuhn to provide the personal perspective he thought necessary to reveal the roots of a paradigm-changing advance in science. And I am immensely gratified by the success of my colleagues and friends worldwide and the strikingly successful progress of the PPRG – an enormously effective knowledge-changing "band of brothers" – who collectively provided the basis for establishment of the new interdisciplinary international fields of Precambrian Paleobiology and Astrobiology. Thanks to these good works, throughout my now nearly 60-year-long quest to provide the answer to Darwin's quandary – the long-sought but previously assumed unknowable so-called "missing" record of Earth's earliest life – virtually every day of my life has been yet Another Day in Paradise.

Addendum

Selected Relevant Literature in Chronological Order

A.1 THE ROOTS OF DARWINIAN EVOLUTION (1796–1813)

Darwin, E. 1796. *Zoonomia: The Laws of Organic Life*, Vols. I and II. London: J. Johnson.

Malthus, T. R. 1798. *Essay on the Principle of Population*. London: J. Johnson.

Lamark, J-B-P-A. De Monet. 1809. *Philosophie Zoologique*. Paris: Muséum d'Histoire Naturelle, Jardin des Plantes.

Cuvier, G. 1813. *Essay on the Theory of the Earth* (1818 English Translation). New York: Kirk and Mercein.

A.2 A CENTURY OF GESTATION (1859–1959)

Darwin, C. R. 1859. *On the Origin of Species by Means of Natural Selection, or the Preservation of Favored Races in the Struggle for Life*. London: John Murray.

Dawson J. W. 1875. *The Dawn of Life, Being the History of the Oldest Known Fossil Remains, and Their Relations to Geological Time and the Development of the Animal Kingdom*. London: Hodder and Stoughton.

Walcott, C. D. 1883. Pre-carboniferous strata of the grand canyon of the Colorado. *American Journal of Science* 26: 437–442.

Walcott, C. D. 1889. Pre-Cambrian fossiliferous formations. *Geological Society of America Bulletin* 10: 199–244.

Walcott, C. D. 1910. Cambrian geology and paleontology II. 1. Abrupt appearance of the Cambrian fauna of the North American Continent. *Smithsonian Miscellaneous Collections* 57: 1–16.

Walcott, C. D. 1915. Discovery of Algonkian bacteria. *Proceedings of the National Academy of Sciences USA* 1: 256–257.

Seward, A. C. 1931. *Plant Life through the Ages, a Geological and Botanical Retrospect*. Cambridge: Cambridge Univ. Press.

Cloud, P. E., Jr. 1942. Notes on stromatolites. *American Journal of Science* 240: 363–379.

Sprigg, R. C. 1947. Early Cambrian jellyfishes from the Flinders Ranges, South Australia. *Transactions of the Royal Society of South Australia* 71: 212–224.

Tyler, S. A. and Barghoorn, E. S. 1954. Barghoorn Occurrence of structurally preserved plants in pre-Cambrian rocks of the Canadian Shield. *Science* 119: 606–608.

Timofeev, B. V. 1957. O novoj gruppe iskopaemykh spor [On a new group of fossil spores]. *Ezhegodnik Vsesoyuznogo Paleontolgiches kogo Obshchestva* 16: 281–285 (in Russian).

Nursall, J. R. 1959. Oxygen as a prerequisite to the origin of the Metazoa. *Nature* 183: 1170–1172.

A.3 BREAKTHROUGH PUBLICATIONS (1961–1965)

Glaessner, M. F. 1961. Pre-Cambrian animals. *Scientific American* 204 (issue 3, March 1961).

Glaessner, M. F. 1962. Pre-Cambrian fossils. *Biological Reviews* 37: 467–494.

Kuhn, T. S. 1962. *The Structure of Scientific Revolutions*. Chicago: Univ. Chicago Press.

Meinschein, W. G., E. S. Barghoorn, and J. W. Schopf. 1964. Biological remnants in Precambrian sediment. *Science* 145: 262–263.

Barghoorn, E. S. and J. W. Schopf. 1965. Microorganisms from the Late Precambrian of central Australia. *Science* 150: 337–339.

Barghoorn, E. S. and S. A. Tyler. 1965. Microorganisms from the Gunflint chert. *Science* 147: 563–577.

Barghoorn, E. S., W. G. Meinschein, and J. W. Schopf. 1965. Paleobiology of Precambrian shale. *Science* 148: 461–472.

Cloud, P. E., Jr. 1965. Significance of the Gunflint (Precambrian) microflora. *Science* 148: 27–35.

A.4 PRECAMBRIAN BIOTAS, STROMATOLITES, AND MOLECULAR BIOMARKERS OF AUSTRALIA, CHINA, INDIA, USA, AND USSR (1968–1988)

Schopf, J. W. 1968, Microflora of the Bitter Springs Formation, Late Precambrian, central Australia. *Journal of Paleontology* 42: 651–688.

Schopf, J. W. and J. M. Blacic. 1971. New microorganisms from the Bitter Springs Formation (late Precambrian) of the north-central Amadeus Basin, Australia. *Journal of Paleontology* 45: 925–960.

Schopf, J. W., T. D. Ford, and W. J. Breed. 1973. Microorganisms from the Late Precambrian of the Grand Canyon, Arizona. *Science* 179: 1319–1321.

Schopf, J. W. and Yu. K. Sovietov. 1976. Microfossils in *Conophyton* from the Soviet Union and their bearing on Precambrian biostratigraphy. *Science* 193: 143–146.

Walter, M. R. (ed.) 1976. *Stromatolites, Developments in Sedimentology 20*. New York: Elsevier.

Schopf, J. W., T. A. Dolnik, I. N. Krylov, *et al.* 1977. Six new stromatolitic microbiotas from the Proterozoic of the Soviet Union. *Precambrian Research* 4: 269–284.

Schopf, J. W. and L. N. Prasad. 1978 Microfossils in *Collenia*-like stromatolites from the Proterozoic Vempalle Formation of the Cuddapah Basin, India. *Precambrian Research* 6: 347–366.

Schopf, J. W. 1982. Current status of Precambrian paleobiology in China: recent discoveries of several new stromatolitic Proterozoic microbiotas. In *Development and Interactions of Precambrian Lithosphere, Biosphere, and Atmosphere*, ed. R. Weber, 33–36. Mexico City: Instituto de Geologia, Universidad Nacional Autonoma de Mexico.

Cloud, P. 1983. Early biogeologic history: the emergence of a paradigm. In *Earth's Earliest Biosphere, Its Origin and Evolution*, ed. J. W. Schopf, 14–31. Princeton: Princeton Univ. Press.

Schopf, J. W., W. Zhu, Z. Xu, and J. Hsu. 1984. Proterozoic stromatolitic microbiotas of the 1,400-1,500 Ma-old Gaoyuzhuang Formation near Jixian, northern China. *Precambrian Research* 24: 335–349.

Summons, R. E., S. C. Brassel, G. Eglington, E. Evans, and R. J. Horodyski. 1988. Distinctive hydrocarbon biomarkers from fossiliferous sediment of the Late Proterozoic Walcott Member, Chuar Group, Grand Canyon, Arizona. *Geochimica Cosmochimica Acta* 52: 2625–2637.

A.5 PRECAMBRIAN PALEOBIOLOGY RESEARCH GROUP (1983–1992)

Schopf, J. W. (ed.). 1983. *Earth's Earliest Biosphere, Its Origin and Evolution*. Princeton: Princeton Univ. Press.

Schopf, J.W. and C. Klein (eds.). 1992. *The Proterozoic Biosphere, A Multidisciplinary Study*. Cambridge: Cambridge Univ. Press.

A.6 MARS "FOSSILS" CONTROVERSY (1996–1997)

McKay, D. S., E. K. Gibson, Jr., K. L. Thomas-Keprta, *et al.* 1996. Search for past life on Mars: possible relic biogenic activity in Martian meteorite ALH84001. *Science* 273: 924–930.

Bradley, J. P., R. P. Harvey, and H. Y. McSween, Jr. 1997. No "nannofossils" in Martian meteorite. *Nature* 390: 454.

A.7 APEX FOSSILS CONTROVERSY (1993–2018)

Schopf, J. W. 1993. Microfossils of the Early Archean Apex chert: new evidence of the antiquity of life. *Science* 260: 640–646.

Brasier, M. D., O. R. Green, A. P., Jephcoat, *et al.* 2002. Questioning the evidence for Earth's oldest fossils. *Nature* 416: 76–81.

García Ruiz, J. M., A. Carnerup, A. G. Christy, N. J. Welham, and S. T. Hyde 2002. Morphology: an ambiguous indicator of biogenicity. *Astrobiology* 2: 353–369.

Schopf, J. W., A. B. Kudryavtsev, D. G. Agresti, T. J. Wdowiak, and A D. Czaja. 2002. Laser-Raman imagery of Earth's earliest fossils. *Nature* 416: 73–76.

Garcia-Ruiz, J. M., S. T. Hyde, A. M. Carnerup, A. G. Christy, M. J. Van Kranendonk, and N. J. Welham. 2003. Self-assembled silica-carbonate structures and detection of ancient microfossils. *Science* 302: 1194–1197.

De Gregorio. B. T., T. G. Sharp, G. J. Flynn, S. Wirick, and R. L. Hervig. 2009. Biogenic origin for Earth's oldest putative microfossils. *Geology* 37: 631–634.

Pinti, D. L., R. Mineau, and V. Clement. 2009. Hydrothermal alteration and microfossil artefacts of the 3,465-million-year-old Apex chert. *Nature Geoscience* 2: 640–643.

Marshall C. P., J. R. Emry, and A. O. Marshall. 2011. Haematite pseudomicrofossils present in the 3.5-billion-year-old Apex chert. *Nature Geoscience* 4: 240–243.

Schopf, J. W. and A. B. Kudryavtsev. 2011. Biogenicity of Apex chert microstructures. *Nature Geoscience* 4: 346–347.

Schopf, J. W. and A. B. Kudryavtsev. 2012. Biogenicity of Earth's earliest fossils: a resolution of the controversy. *Gondwana Research* 39: 61–771.

Brasier, M. D., J. Antcliffe, M. Saunders, and D. Wacey. 2015. Changing the picture of Earth's earliest fossils (3.5–1.9 Ga) with new approaches and new discoveries. *Proceedings of the National Academy of Sciences USA* 112: 4859–4864.

Wacey, D., M. Saunders, C. Kong, A. Brasier, and M. Brasier. 2016. 3.46 Ga Apex chert "microfossils" reinterpreted as mineral artifacts produced during phyllosilicate exfoliation. *Gondwana Research* 36: 296–313.

Schopf, J. W., K. Kitajima, M. J. Spicuzza, A. B. Kudryavtsev, and J. W. Valley. 2018. SIMS analyses of the oldest known assemblage of microfossils document their taxon-correlated carbon isotope compositions. *Proceedings of the National Academy of Sciences USA* 115: 53–58.

A.8 NEW ANALYTICAL TECHNIQUES (2000–2010)

House, C. H., J. W. Schopf, K. D. McKeegan, C. D. Coath, T. M. Harrison, and K. O. Stetter. 2000. Carbon isotopic composition of individual Precambrian microfossils. *Geology* 28: 707–710.

Kudryavtsev, A. B., J. W. Schopf, D. G. Agresti, and T. J. Wdowiak. 2001. *In Situ* laser-Raman imagery of Precambrian microscopic fossils. *Proceedings of the National Academy of Sciences USA* 98: 823–826.

Schopf, J. W. and A. B. Kudryavtsev. 2005. Three-dimensional Raman imagery of Precambrian microscopic organisms. *Geobiology* 3: 1–12.

Schopf, J. W., A. B. Kudryavtsev, D. G. Agresti, A. D. Czaja, and T. J. Wdowiak. 2005. Raman imagery: a new approach to assess the geochemical maturity and biogenicity of permineralized Precambrian fossils. *Astrobiology* 5: 333–371.

Schopf, J. W., A. Tripathi, and A. B. Kudryavtsev. 2006. Three-dimensional confocal optical microscopy of Precambrian microscopic organisms. *Astrobiology* 6: 1–16.

Schopf, J. W. and A. B. Kudryavtsev. 2009. Confocal laser scanning microscopy and Raman imagery of ancient microscopic fossils. *Precambrian Research* 173: 39–49

Schopf, J .W. and A. B. Kudryavtsev. 2010. A renaissance in studies of ancient life. *Geology Today* 26: 141–146.

A.9 RECENT ADVANCES (2015–2018)

Schopf, J. W., A. B. Kudryavtsev, M. R. Walter, *et al.* 2015. Sulfur-cycling fossil bacteria from the 1.8-Ga Duck Creek Formation provide promising evidence of evolution's null hypothesis. *Proceedings of the National Academy of Sciences USA* 112: 2087–2092.

Garcia, A. K., J. W. Schopf, S.-I. Yokobori, S. Akanuma, and A. Yamagishi. 2017. Reconstructed ancestral enzymes suggest long-term cooling of Earth's photic zone since the Archean. *Proceedings of the National Academy of Sciences USA*. 114: 4619–4624.

Schopf, J. W., A. B. Kudryavtsev, J. T. Osterhout, *et al.* 2017. An anaerobic ~3400 Ma shallow-water microbial consortium: presumptive evidence of Earth's Paleoarchean anoxic atmosphere. *Precambrian Research* 299: 309–318.

Schopf, J. W., K. Kitajima, M. J. Spicuzza, A. B. Kudryavtsev, and J. W. Valley. 2018. SIMS analyses of the oldest known assemblage of microfossils document their taxon-correlated carbon isotope compositions. *Proceedings of the National Academy of Sciences USA* 115: 53–58.

Index